SpringerBriefs in Electrical and Computer Engineering

Series Editors

Woon-Seng Gan, School of Electrical and Electronic Engineering, Nanyang Technological University, Singapore, Singapore

C.-C. Jay Kuo, University of Southern California, Los Angeles, CA, USA

Thomas Fang Zheng, Research Institute of Information Technology, Tsinghua University, Beijing, China

Mauro Barni, Department of Information Engineering and Mathematics, University of Siena, Siena, Italy

T0172180

SpringerBriefs present concise summaries of cutting-edge research and practical applications across a wide spectrum of fields. Featuring compact volumes of 50 to 125 pages, the series covers a range of content from professional to academic. Typical topics might include: timely report of state-of-the art analytical techniques, a bridge between new research results, as published in journal articles, and a contextual literature review, a snapshot of a hot or emerging topic, an in-depth case study or clinical example and a presentation of core concepts that students must understand in order to make independent contributions.

More information about this series at http://www.springer.com/series/10059

János Ladvánszky

Theory of Power Matching

 Springer

János Ladvánszky
Ericsson Telecom Hungary
Budapest, Hungary

ISSN 2191-8112   ISSN 2191-8120 (electronic)
SpringerBriefs in Electrical and Computer Engineering
ISBN 978-3-030-16630-4   ISBN 978-3-030-16631-1 (eBook)
https://doi.org/10.1007/978-3-030-16631-1

This Springer imprint is published by the registered company Springer Nature Switzerland AG.
The registered company address is: Gewerbestrasse 11, 6330 Cham, Switzerland

To Mrs. Gabriella Reeves, My Cello Teacher

Preface

The reader has been invited now for a grandiose trip. One of the most important concepts during the history of circuit theory will be overviewed: power matching. This concept has a great impact on various areas such as foundation, broadband matching, and game theory.

The eBook consists of 12 chapters. We present an overview in Chap. 1. In Chap. 2, a completely derivative-free approach is shown for proving the globality of the maximum power in case of linear time invariant one-ports. In Chap. 3, nonlinear resistive case has been treated: the well-known solar cell example. Topic of Chap. 4 is competitive power matching. That means a multiport case when the ports have been optimized consecutively. This way leads to very different solution than the simultaneous optimization. The big Chap. 5 deals with the scattering matrix as the measure from the power matched status. Broadband matching applications have been touched as classical and real frequency methods for broadband matching. In Chap. 6, we prove that a linear, passive one-port is always causal and that causality can be hurt in case of large nonlinear circuits. Chap. 7 is intended for our own results: power matching with describing functions and weakly nonlinear circuits . In Chap. 8, the most general case has been solved: nonlinear dynamic sources. Importance of the Gateaux derivative has been pointed out. This work has been concluded in Chap. 9.

Technically, main units are called as chapters, while sub-chapters are called sections. Figures and equations are numbered by the chapter and section number and the consecutive serial number. In the references chapter, literature has been grouped according to the chapters.

For those who are less familiar with circuit theory, we suggest studying the references for this chapter before reading this work. Suggested order of reading is: [Dieudonné60, Gantmacher59, Papoulis77, Valkenburg74, Belevitch68, Chua87].

Budapest, Hungary   János Ladvánszky

References

[Belevitch68] V. Belevitch, *Classical Network Theory* (Holden-Day, 1968)

[Chua87] L.O. Chua, C.A. Desoer, E.S. Kuh, *Linear and Nonlinear Circuits* (McGraw-Hill, 1987)

[Dieudonné60] J. Dieudonné, *Foundations of Modern Analysis* (Academic Press, New York, 1960)

[Gantmacher59] F.R. Gantmacher, *The Theory of Matrices* (Chelsea, 1959)

[Papoulis77] A. Papoulis, *Signal Analysis* (McGraw-Hill, 1977)

[Valkenburg74] M.E. Van Valkenburg, *Circuit Theory: Foundations and Classical Contributions* (Dowden, Hutchinson & Ross, 1974)

Acknowledgments

The three persons who influenced this work the most are Dr. A. Baranyi, Professor T. Berceli, and Dr. E. K. Simonyi. Dr. A. Baranyi was my first master. My first work in this topic was supervised by him: the reversal of the nonlinear resistive maximum power theorem. Later I learnt microwaves and optics from Professor T. Berceli. More recently, he sent talented students to me whom I could consult in their diploma works. Dr. E. K. Simonyi was the advisor of my D.Sc. work. He combined rigorous criticism with friendly approach.

Basic influences were made by my teachers at the Budapest University of Technology and Economics: Professor K. Simonyi, Professor L. Pap, Professor V. Székely, Professor S. Csibi, and Professor K. Géher. I learnt from them the basics of electrical engineering. Without this knowledge, this eBook could not be imagined.

In 1976, one of my problems at the closing exam from Electromagnetic Theory was power matching. I admit it is not intentional that the present eBook has been written in the same topic. I would like to mention my examiners, Professor Gy. Veszely, still active, and Dr. S. Takács. I still remember the constructive atmosphere of this exam. Dr. S. Takács taught the subject Electromagnetic Theory to our university group, and I clearly remember his thorough explanations and his beautiful handwriting on the blackboard.

It is my special honor mentioning here the long-term communications at conferences with Professor L. O. Chua whose works are frequently referred here. He is among the professionals in electrical engineering to whom I look up.

From the history of this eBook, personal meeting with Professor M. Hasler at the European Conference on Circuit Theory and Design in 1987 (and several times since then) was significant. Professor M. Hasler was one of the reviewers of the work about power matching using my describing functions. The encouragement that I received from him helped my work in this topic very much.

In this eBook, the computer program AWR is frequently used. The parent of AWR is APLAC by Professor M. Valtonen. My pleasant stay with him and his group at Helsinki University of Technology in 1992, now Aalto University, was a significant event in my career.

Describing function approach to the problem of power matching was a part of my Ph.D. thesis. Now, the opponents of this work are acknowledged, Professor I. Frigyes and Dr. I. Schmideg. Their constructive criticisms and long-term communications with them helped me a lot.

When this eBook was born I imagined that I explained to my students or young colleagues. I am grateful to all of them for keeping me mentally young: Zoltán Szalontai, Dr. Attila Hilt, Dr. Gábor Kovács, Gergely Mészáros, Bence Erdei, Tibor Gál, Kristóf Máté Osbáth, András Surman, Tamás Szili, Lajos Budai, and Ádám Kristóf Ónodi.

It is my special honor mentioning my friends (who are great professionals at the same time) who helped me in refinements: Dr. E. K. Simonyi, Dr. A. Somogyi, Mr. Z. Huszka, Dr. A. Hilt, and Dr. A. Zólomy.

Big thanks are due to Dr. B. Kovács, my boss at Ericsson Hungary, who made it possible for me to work at home before retirement.

This research was supported by the National Research Development and Innovation Office of Hungary within the Quantum Technology National Excellence Program (Project No. 2017-1.2.1-NKP-2017-00001).

Abstract

Power matching is one of the most important concepts in the history of circuits and systems theory. In this eBook, the most significant results and the impact of power matching on other concepts have been presented.

Reasons of writing about power matching are the rich impact on other areas and the vast amount of applications. From impacts, basis of today's microwave industry, the scattering matrix should be emphasized first. The scattering matrix also appears in digital circuits, for example, the wave digital filters to be included in the next edition. As far as the applications are considered, our results in describing functions and Volterra series are mentioned. From this sample palette, applications in numerical solution for the Nash equilibrium should not be left out, which is intended to be detailed later. The most distant application that we can still see is quantum communications, also detailed in a latter edition.

Contents

About the Author

János Ladvánszky (member, IEEE) is an electrical engineer at Ericsson Hungary (M.Sc. in 1978, "Modelling of a microwave transistor," from the Budapest University of Technology and Economics; Ph.D. in 1988, "Nonlinear, microwave circuit design," from the Hungarian Academy of Sciences; D.Sc. defense, forthcoming, "Integrated systems for optical communications," at the Hungarian Academy of Sciences). His career can be followed at ResearchGate, LinkedIn, and Facebook. He has got 180 conference and journal papers and 14 patents, 52 independent citations, and above 3400 reads at ResearchGate.

Chapter 1
Overview

1.1 From Linear to Nonlinear Circuits

Way of the solution led us historically through the generalization of the conjugate matching condition for multiports, to generalization for nonlinear generators and through solution of special power matching problems, as it will be detailed in the following. The solution of the power matching problem serves for the introduction of the scattering matrix, its extension for the entire complex frequency plane and founding for the broadband matching problem, whose generalization is, under the name of interpolation by positive real matrices, known from many related research areas. The roots of this problem can be found within the foundations of circuit theory [Youla59], namely, by presenting the relation between passivity and causality of linear circuits.

The earliest publications in power matching can be found between references by [Belevitch48]. Accordingly, the problem appeared in the 1920s, in connection with telegraph networks. This is not detailed here; instead, we start with being aware of the solution for linear time invariant one-ports, at one frequency point.

[Baudrand70] deals with power matching of linear, complex, nonreciprocal multiports. Extremum of the power has been found by varying the load impedance matrix, and consequently the current variations. The conclusion has been that for the existence of the power maximum, impedance matrix of the generator should be positive definite. The result has been that for power maximum, load impedance matrix is the conjugate transpose of the generator impedance matrix. At the end of the paper, the case of the uncoupled loads has also been considered, for fixed generator voltages.

[Spinei72] continues that Baudrand thinks and underlines the importance of passivity of the generator impedance. He investigates fixed generator voltages. It has been shown there are infinitely many loads dissipating maximum power. It has been proved that in case of extremum power, total generator power is divided equally between the generator impedance matrix and the loads.

© The Author(s), under exclusive licence to Springer Nature Switzerland AG 2019
J. Ladvánszky, *Theory of Power Matching*, SpringerBriefs in Electrical and
Computer Engineering, https://doi.org/10.1007/978-3-030-16631-1_1

[Mathis72] outlines the possibility of maximizing the load power by a 1-Ω resistor connected to a lossless circuit. In his example, the lossless circuit is a transformer comprising n primer and one seconder coils that supplies the whole power to the 1-Ω resistor.

[Lin72] extends the concept of maximum available power for the case of multiport generators. Resistive generators have been investigated and loaded by uncoupled resistors. He points out that expressing the necessary condition of the power maximum by differentiating the total power with respect to the load resistors, the equation is nonlinear and this way the solution is difficult. He also points out that this is only a necessary condition and it requires further investigations. Then, applying the first variation of the load resistance matrix, an equation has been obtained for the load resistances as unknowns, and its solution has been discussed.

An essential simplification and clarification has been obtained by [Desoer73]. With respect to those works cited by him, the main change is that in power matching, not the load impedance (impedance matrix) but the load currents are the independent variables. Thus, the problem has been divided into two parts: finding the currents first and then the impedance matrix. He shows that for fixed generator voltages, the solution is not unique; for n-ports, it is a $n(n-1)$ dimensional submanifold of the field of complex numbers. He shows that for arbitrary generator voltages, the solution becomes unique; in this case, load impedance matrix is the conjugate transpose of the generator impedance matrix. He points out that for the existence of the solution it is not necessary to assume passivity of the load impedance matrix because this is a consequence. Degenerate cases have also been discussed.

It appeared earlier in time, but this is the proper place to mention [Rohrer65] who looks for the conditions of maximum total extracted energy from a linear time-variant generator with arbitrary waveform. He was the first recognizing that instead of the n^2 parameters of the load, it is simpler to vary n load currents. He shows that the solution consists of fulfilling two conditions called as adjoint matching and energy absorbing conditions. Adjoint matching is a necessary condition: Weight function matrix of the optimum load is conjugate transpose of that of the load, changing the sign of the time variables. The energy absorbing condition is sufficient for maximum energy. Accordingly, weight function matrix of the generator is positive definite. Then the formalism has been applied for linear, time-invariant generators in order to introduce complex normalized scattering matrix as a measure of deviation from the matched case. Many elements of these investigations have also been found in his later work [Rohrer68] in which the concept of networks composed by complex elements has been avoided.

[Vidyasagar74] follows the line by Desoer. He underlines that although it is not necessary to assume load passivity, but possible. He seeks a load impedance matrix satisfying the conditions for solvability, passivity, and power optimum, for fixed generator voltages. He shows there are infinitely many such possibilities and shows a method for generating them.

[Flanders76] obtains a unified treatment of known results from the viewpoint of mathematics.

[Calvaer83] recognizes that the condition for generator impedance matrix has not been satisfied in power electronics, but power maximum still exists. Purely inductive generator has been investigated, or in general, purely imaginary, symmetrical, positive definite generator impedance matrix. In this case, impedance matrix of the maximum power load is identical to the reactance matrix of the generator. This load is called as mirror circuit, and he shows that mirror circuit load is necessary and sufficient for the power maximum.

The next significant paper is [Desoer83], who strengthens Calvaer's results. The mirror circuit is the only load that absorbs maximum power at arbitrary generator voltages. Calvaer was not clear in saying whether the maximum is local or global. Desoer clarifies: The maximum that is reached by mirror circuit load is global. He points out the special interest of the solution: Analogously with the generators of positive definite impedance matrix, the result is not unique but a $n(n - 1)$ dimensional submanifold of the complex numbers.

[Wyatt83B] treats power matching of nonlinear resistive generators. This is the first paper about nonlinear version of the problem with circuit theoretical rigor. Assuming the Norton equivalent of the generator, he obtains closed form characteristics of the load dissipating maximum power. The result is significant because it is related to power matching of solar cells, and this problem was previously solved in practice by applying a complicated maximum power tracker that finds adaptively the optimum operating point. The authors underline that these complicated circuits are not necessary in theory. In practice, maximum power trackers may be necessary due to aging and temperature dependence of the solar cells, although the secondary effects result in only a small deviation from the theoretically given optimum operating point.

[Lin85] turns again to linear generators. He deals with a power matching problem motivated by the analogy between electrical circuits and economical processes. We have a linear two-port generator that is resistive. We connect to the generator ports, linear, purely resistive, uncoupled, variable loads. Load resistors are varied alternatively, so that at every step, power of the varied load is maximum. The problem is, determining and characterizing the steady state. The problem is called as competitive power matching, in contrary with cooperative power matching that is maximization of the total power. He proves that the solutions are the image impedances corresponding to the generator impedance (admittance) matrix. He also proves that the solutions are the absolute values of the image impedances if the generator is purely reactive. He guesses that the solution is also valid for n-ports, but in this case, it is difficult to determine the image impedances. His interesting example is power matching of a generator whose impedance matrix contains 1-s at all entries. In this case, cooperative power matching solution is a finite power value and that of competitive power matching, tends to zero at both ports.

[Wyatt88, as a research report, Wyatt83A] goes the furthest who formulates the maximum power theorem, for strongly nonlinear, dynamic generators and for arbitrary waveforms. The result has similar form than that for nonlinear, resistive generators, but the usual derivative is replaced by (Gateaux) operator derivative. The theorem contains the solvability of the equations describing the generator-load pair, and the condition for the unique solution. The solution is based on the existence of the Norton equivalent of the generator. Referring to our work [Ladvanszky85a] he notes that although this is the initial condition, nothing can be said in general about its fulfillment. We note that although his one is the most general solution of the problem, its practical application is difficult for two reasons. One is that the characteristics of the optimum load are difficult to realize due to the present operator derivative. The other reason is that often the solution leads to noncausal circuits.

The practically important cases [Ladvanszky87a, Ladvanszky88c] appeared independently from Wyatt's work, using different model, although formally they can be considered as special cases of [Wyatt88].

1.2   Describing Functions

Reaching maximum output power of microwave power amplifiers and other circuits such as oscillators, frequency multipliers is an old topic in the repertoire of researchers. This is a living problem, which is proved by the large number of publications [Vendelin78, Tserng79, Tucker80, Honjo81, Curtice85, etc.]. The problem includes modelling, measurement, and circuit design parts. In this chapter, we deal with the circuit design problem how we can calculate the characteristics of the maximum power load based on the previously determined model. Thus, we only touch the modelling and measurement part problems and we concentrate on the circuit design.

In modelling bandpass circuits, it is widespread using the describing function method. Traditionally, describing function has been defined as the ratio of the complex amplitudes of the first harmonic component of the answer for sinusoidal excitation and the excitation itself. [Gelb68] showed that this is a much more general problem. The describing functions regarding to different excitation shapes have been defined as transfer functions of group of filters that approximate the output signal in least squared error sense. This approach offers approximate modelling of nonlinear elements excited by different signals or by their sum, and thus in addition to sinusoidal excited describing functions, definitions of other describing functions have also been obtained. In this chapter we shall apply only the sinusoidal excited describing functions, but the mentioned approach gives broader foundation and inspiration for generalization.

For modelling of nonlinear microwave two-ports, basically three different methods have been applied. Previously an extension for the linear scattering matrix was suggested in which the entries of the matrix are assumed level-dependent [Chaffin73, Soares77, Mazumder78], etc. Applicability of these "large-signal S-parameters" has been severely limited, however.

The method ("load-pull") in which the device has been characterized by the locus of output reflection for a fixed input power is applicable for solving practical problems [Cusack74, Takayama76, Zemack80, Poulin80, Tucker84]. This method has been applied for the design of amplifiers that do not contain feedback [Tserng79, Arai79, Honjo81, Curtice85]. Disadvantage is that a rather complicated and expensive measurement is necessary and cannot be applied when feedback cannot be neglected.

This latter disadvantage is eliminated by the introduction of two-port describing functions [Mazumder77, Baranyi86], but this method leads to further increase of data amount.

One of the goals of all three methods is to determine the load reflection corresponding to maximum output power. But for the question what the relation of the optimum load to the model parameters of the device is, there is no clear and unique answer.

As an influence of the statements from linear circuits, it has been assumed for a long time that conjugate of the large signal output reflection gives the maximum power [Rauscher80]. The earliest note that this is not so can be found in [Vendelin78]. [Tucker79] has a conclusion from measurement data that with increasing the input power, optimum load reflection deviates more and more from the conjugate of small signal reflection. In a later publication [Tucker80], an attempt has been made for determining the relation between large signal output reflection and optimum load. He assumed that the output of the investigated nonlinear active two-port can be modelled by a current source in parallel with a nonlinear conductance and a capacitance. He gives an empirical relation between the generator and optimum load conductances. In a later publication [Tucker81], the capacitance has been assumed linear. His statement is that optimum load susceptance is the negative of the generator susceptance.

Most general solution for power matching of nonlinear, dynamic n-ports has been obtained by [Wyatt83]. But the author notes that the optimum load given by him is noncausal in general, and thus these results cannot be applied in practice.

It is obvious that at this point, there was a wide gap between general and practically applicable results. We filled this gap in Chap. 7. Nonlinear resistive maximum power theorem has been extended for the case of nonlinear, dynamic generators characterized by describing functions. We show how linear results can be deduced from our ones. We determine optimum load in terms of admittance and scattering describing functions as well. The latter relation has been verified by computer analysis.

1.3   Weakly Nonlinear Circuits

Weakly means here polynomial nonlinearity for resistive circuits, which is generalized as simple, double, and triple (and so on) convolutions for non-resistive circuits. This latter representation is called as Volterra series. Originally it was found as a tool for solving integro-differential equations [Volterra27]. A recent overview is [Franz06]. In electrical engineering, it has been rediscovered by Chua and his colleagues [Chua79, Boyd84]. Volterra series are especially useful for analysis of nonlinear circuits excited by modulated signals [Kuo77]. Mathematical properties of Volterra series have been studied in numerous papers by Sandberg [Sandberg].

As I know the problem of power matching using Volterra series was intact before my first publication in this topic.

Finally we mention our works in the field of nonlinear power matching [Ladvanszky85b, Ladvanszky86a, Ladvanszky86b, Ladvanszky87b, Ladvanszky88a, Ladvanszky88b, Ladvanszky89, Ladvanszky99].

References

International References

[Arai79]   Y. Arai et al., High power GaAs FET amplifier for TWT replacement. Fujitsu
                Sci. Tech. J. **15**(3), 63–82 (1979)
[Baranyi86]   A. Baranyi, Modelling of large-signal microwave devices, in Hungarian,
                Híradástechnika, 6/1986, pp. 273-280
[Belevitch48]   V. Belevitch, Transmission losses in 2n-terminal networks. J. Appl. Phys. **19**(7),
                636–638 (1948). in *Circuit Theory: Foundations and Classical Contribu-
                tions*, ed. by M.E. Van Valkenburg (Dowden, Hutchinson and Ross, Inc.,
                1974)
[Baudrand70]   H. Baudrand, On the generalizations of the maximum power transfer theorem.
                Proc. IEEE **58**, 1780–1781 (1970)
[Boyd84]   S. Boyd, L.O. Chua, C.A. Desoer, Analytical foundations of Volterra series.
                IMA J. Math. Control Info. **1**, 243–282 (1984)
[Calvaer83]   A.J. Calvaer, On the maximum loading of active linear electric multiports. Proc.
                IEEE **71**, 282–283 (1983)
[Chaffin73]   R.J. Chaffin, W.H. Leighton, Large-signal S-parameter characterization of
                UHF power transistors, in *IEEE MTT International Microwave Symposium,
                Dig. Tech. Papers, University of Colorado, Boulder, CO*, (5 June 1973),
                pp. 155–157
[Chua79]   L.O. Chua, C.Y. Ng, Frequency domain analysis of nonlinear systems: general
                theory. Electron. Circuits Syst., 165–185 (1979). https://doi.org/10.1049/ij-
                ecs:19790030
[Curtice85]   W.R. Curtice, M. Ettenberg, A nonlinear GaAs FET model for use in the design
                of output circuits for power amplifiers. IEEE Trans. Microwave Theory
                Tech. **33**(12), 1383–1394 (1985)
[Cusack74]   J.M. Cusack et al., Automatic load contour mapping for microwave power
                transistors. IEEE Trans. Microwave Theory Tech. **22**(12), 1146–1152 (1974)
[Desoer73]   C. A. Desoer: „The maximum power transfer theorem for n-ports", IEEE Trans.
                Circuit Theory, 1973, 20328–330
[Desoer83]   C.A. Desoer, A maximum power transfer problem. IEEE Trans. Circuits Syst.
                CAS-30(10), 757–758 (1983)
[Flanders76]   H. Flanders, On the maximal power transfer theorem for n-ports. Circuit Theory
                Appl **4**, 319–344 (1976)
[Franz06]   M.O. Franz, B. Schölkopf, A unifying view of Wiener and Volterra theory and
                polynomial kernel regression. Neural Comput. **18**(12), 3097–3118 (2006)
[Gelb68]   A. Gelb, W.E. Vander Velde, *Multiple-input describing functions and nonlinear
                system design* (McGraw-Hill, New York, 1968)
[Honjo81]   K. Honjo, Y. Takayama, A 25W 5GHz GaAs FET amplifier for a microwave
                landing system. IEEE Trans. Microwave Theory Tech. **29**(6), 579–582
                (1981)
[Kuo77]   Y.L. Kuo, Frequency-domain analysis of weakly nonlinear networks. IEEE
                Trans. Circuits Syst.. **CS-11**(4) 1977; **CS-11**(5), 2–6 (1977)
[Lin72]   P.M. Lin, Determination of available power from resistive multiports. IEEE
                Trans. Circuit Theory **19**, 385–386 (1972)

[Lin85]	P.M. Lin, Competitive power extraction from linear n-ports. IEEE Trans. Circuits Syst. **32**, 185–191 (1985)
[Mathis72]	B.A. Mathis, H.F. Mathis, Maximum power transfer from a multiple-terminal network to a single impedance. Proc. IEEE **60**, 746 (1972)
[Mazumder77]	S.R. Mazumder, P.D. Van der Puije, An experimental method of characterizing nonlinear two-ports and its application to microwave class C transistor power amplifier design. IEEE J. Solid State Circuits **12**(5), 576–580 (1977)
[Mazumder78]	S.R. Mazumder, P.D. Van der Puije, Two-signal method of measuring the large signal S-parameters of transistors. IEEE Trans. Microwave Theory Tech. **26**(6), 417–420 (1978)
[Poulin80]	D. Poulin, Load-pull measurements help you meet your match. Microwaves **19**, 61–65 (1980)
[Rauscher80]	C. Rauscher, H.A. Willing, Design of broad-band GaAs FET power amplifiers. IEEE Trans. Microwave Theory Tech. **28**(10), 1054–1059 (1980)
[Rohrer65]	R.A. Rohrer, The scattering matrix normalized to complex n-port load networks. IEEE Trans. Circuit Theory **12**, 223–230 (1965)
[Rohrer68]	R.A. Rohrer, Optimal matching: a new approach to the matching problem for real time-invariant one-port networks. IEEE Trans. Circuit Theory **15**, 118–124 (1968)
[Sandberg]	I.W. Sandberg, Multidimensional nonlinear myopic maps, Volterra series, and uniform neural-network approximations, in *Intelligent Methods in Signal Processing and Communications*, ed. by D. Docampo et al., (Birkhäuser, Boston, 1997)
[Soares77]	R.A. Soares, Novel large signal S-parameter measurement technique aids GaAs power amplifier design, in *Proceedings of 6th EuMC*, (September 1977), pp. 113–117
[Spinei72]	F. Spinei, On generalizations of the maximum power transfer problem. Proc. IEEE **60**, 903–904 (1972)
[Takayama76]	Y. Takayama. A new load-pull characterization method for microwave power transistors, in *International Microwave Symposium,* Dig. Tech. Papers, June 1976, pp. 218–220
[Tserng79]	H.Q. Tserng, Design and performance of microwave power GaAs FET amplifiers. Microw. J **22**(6), 94–100 (1979)
[Tucker79]	R.S. Tucker, Third order intermodulation distortion and gain compression in GaAs FET's. IEEE Trans. Microwave Theory Tech. **27**(5), 400–408 (1979)
[Tucker80]	R.S. Tucker, Optimum load admittance for a microwave power transistor. Proc. IEEE **68**(3), 410–411 (1980)
[Tucker81]	R.S. Tucker, RF characterization of microwave power FETs. IEEE Trans. Microwave Theory Tech. **29**(8), 776–781 (1981)
[Tucker84]	R.S. Tucker, P.D. Bradley, Computer-aided error correction of large-signal load-pull measurements. IEEE Trans. Microwave Theory Tech. **32**(3), 296–300 (1984)
[Vendelin78]	G.D. Vendelin, Power GaAs FET amplifier design with large signal tuning parameters, in *IEEE 1978 Asilomar Conference on Circuits and Systems*, Dig. Tech. Papers, Pacific Grove, CA (1978), pp. 139–141
[Vidyasagar74]	M. Vidyasagar, Maximum power transfer in n ports with passive loads. IEEE Trans. Circuits Syst. **CAS-21**(3), 327–330 (1974)
[Volterra27]	V. Volterra, Theory of functionals and of integrals and integro-differential equations. Madrid 1927 (Spanish), translated version reprinted Dover Publications, New York (1959)
[Wyatt83]	J.L. Wyatt. Nonlinear dynamic maximum power theorem, with numerical method, Research Report, MIT, LIDS-P-1331, September 1983
[Wyatt83A]	J.L. Wyatt, Nonlinear dynamic maximum power theorem, with numerical method. Internal report, Massachusetts Institute of Technology, LIDS-P-1331 (1983)

8   1   Overview

[Wyatt83B]   J.L. Wyatt, L.O. Chua, Nonlinear resistive maximum power theorem, with solar
              cell application. IEEE Trans. Circuit Syst. **30**, 824–828 (1983)
[Wyatt88]   J.L. Wyatt, Nonlinear dynamic maximum power theorem. IEEE Trans. Circuits
              and Syst. **35**(5), 563–566 (1988)
[Youla59]   D.C. Youla, L.J. Castriota, H.J. Carlin, Bounded Real scattering matrices and
              the foundation of linear passive network theory. IRE Trans. Circuit Theory
              CT-6(1), 102–124 (1959)
[Zemack80]   D. Zemack, A new load-pull measurement technique eases GaAs characteriza-
              tion. Microw. J. **23**(11), 63–67 (1980)

Own Publications

[Ladvanszky85a]   J. Ladvánszky, A. Baranyi, On power matching of nonlinear resistive sources.
                  Proceedings of the European Conference on Circuit Theory and Design.
                  ECCTD'85, Prague, Czechoslovakia, 2–6 September 1985, pp. 186–188
[Ladvanszky85b]   J. Ladvánszky, Nemlineáris rezisztív áramkörök teljesítmény-illesztése
                  (power matching of nonlinear resistive circuits, in Hungarian), a TKI
                  Közleményei, Budapest, 1985/3–4, 53–70. old
[Ladvanszky86a]   J. Ladvánszky, On the extension of the nonlinear resistive maximum power
                  theorem I. Proceedings of the International Symposium on Circuits and
                  Systems, ISCAS'86, San José, California, USA, May 5–7, pp. 257–259,
                  (1986)
[Ladvanszky86b]   J. Ladvánszky, On the extension of the nonlinear resistive maximum power
                  theorem II. Proceedings of the International Colloquium on Microwave
                  Communications, Budapest, Hungary, Aug. 25–29, pp. 251–252 (1986)
[Ladvanszky87a]   J. Ladvánszky, Maximum power theorem - a describing function approach.
                  Proceedings of the European Conference on Circuit Theory and Design,
                  ECCTD'87, Paris, France, September 1–4, pp. 35–40 (1987)
[Ladvanszky87b]   J. Ladvánszky, Teljesítmény-maximalizálás a leírófüggvény-módszer
                  alkalmazásával (Power maximization applying describing function
                  approach, in Hungarian). a TKI Közleményei, Budapest, 1987/2., 61–80.
                  old
[Ladvanszky88a]   J. Ladvánszky, Nemlineáris, mikrohullámú áramkörök tervezésének
                  problémái: teljesítményillesztés, a reflexiós mátrix mérési hibáinak
                  korrekciója (Some problems of nonlinear, microwave circuit design, in
                  Hungarian). kandidátusi értekezés, Magyar Tudományos Akadémia
                  (1988) március 7
[Ladvanszky88b]   J. Ladvánszky, Nemlineáris, mikrohullámú áramkörök hasznos
                  teljesítményének maximalizálása (Maximizing effective power in
                  nonlinear, microwave circuits, in Hungarian). a Bognár Géza Emlékülés
                  kiadványa (Proceedings of the memorial session in honor of Géza Bognár),
                  Budapest (1988). április 20–21., 125–130. old
[Ladvanszky88c]   J. Ladvánszky, Maximum power transfer in weakly nonlinear circuits. Pro-
                  ceedings of the International Symposium on Circuits and Systems,
                  ISCAS'88, Helsinki, Finland, June 7–9, pp. 2723–2726 (1988)
[Ladvanszky89]   J. Ladvánszky, Nemlineáris, mikrohullámú áramkörök teljesítményillesztése
                  (Power matching in nonlinear, microwave circuits, in Hungarian).
                  Híradástechnika, 1989/3., 89–95. old
[Ladvanszky99]   J. Ladvánszky, Maximális teljesítmény-átvitel kis nemlinearitású
                  áramkörökben (Maximum power transfer in weakly nonlinear circuits, in
                  Hungarian). Híradástechnika, 1999/6. 8–12. old

Chapter 2
Linear, Time Invariant One Ports: A Derivative-Free Proof of the Global Optimum

Theorem 2.1 Let i and v denote the complex current and voltage of the generator port in the frequency domain, respectively. Then we state that the generator power

$$P = \text{Re}(i^*v) \tag{2.1}$$

where the star denotes complex conjugate, takes its global maximum for arbitrary source current i_S when for the load admittance, the following holds:

$$Y_L = Y_S^* \tag{2.2}$$

provided that the source admittance has a positive real part.

Proof According to Fig. 2.1,

$$v = \frac{i_S}{Y_S + Y_L} \tag{2.3}$$

$$i = Y_L v \tag{2.4}$$

Thus

$$P = \text{Re}\left(Y_L^* \left| \frac{i_S}{Y_S + Y_L} \right|^2 \right) \tag{2.5}$$

© The Author(s), under exclusive licence to Springer Nature Switzerland AG 2019
J. Ladvánszky, *Theory of Power Matching*, SpringerBriefs in Electrical and
Computer Engineering, https://doi.org/10.1007/978-3-030-16631-1_2

Fig. 2.1 The linear time invariant generator and the load

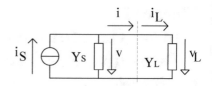

We state that the maximum power is

$$P_{\max} = \mathrm{Re}\left(Y_S^* \left| \frac{i_S}{Y_S + Y_S^*} \right|^2 \right) \tag{2.6}$$

and now we prove that

$$P_{\max} - P \geq 0 \tag{2.7}$$

Substituting (2.5, 2.6) into (2.7):

$$P_{\max} - P = G_S \left| \frac{i_S}{Y_S + Y_S^*} \right|^2 - G_L \left| \frac{i_S}{Y_S + Y_L} \right|^2 \tag{2.8}$$

Rearrangement of (2.8) yields

$$
\begin{aligned}
P_{\max} - P &= |i_S|^2 \frac{G_S\left[|Y_S|^2 + Y_S Y_L^* + Y_L Y_S^* + |Y_L|^2\right] - G_L 4 G_S^2}{|Y_S + Y_L|^2 |Y_S + Y_S^*|^2} \\
&= |i_S|^2 G_S \frac{\left[|Y_S|^2 - 2G_S G_L - 2B_S B_L + |Y_L|^2\right]}{|Y_S + Y_L|^2 |Y_S + Y_S^*|^2} \\
&= |i_S|^2 G_S \frac{|Y_S - Y_L^*|^2}{|Y_S + Y_L|^2 |Y_S + Y_S^*|^2}
\end{aligned} \tag{2.9}
$$

From (2.9) we can see that (2.7) always holds, with equality if and only if (2.2) is satisfied (conjugate matching).

The physical meaning of (2.9) is that there is a unique global maximum, at conjugate matching.

Chapter 3
Nonlinear, Resistive Case

3.1 Power Matching of a Solar Cell (Wyatt, Chua)

Model of a loaded solar cell is shown in Fig. 3.1.1.

The problem is as follows. Given the source characteristics $f(v)$, it is to find the nonlinear load characteristics that results in absorbing maximum possible power at arbitrary value of i_S.

The source characteristic $f(v)$ must obey some conditions; this will be treated later, in Sect. 3.2.

The simplest solution is based on the derivative of the extracted power P:

$$P = i_L v_L \qquad (3.1.1)$$

where the load characteristics to be determined is $g(v)$:

$$i_L = g(v_L) \qquad (3.1.2)$$

From Fig. 3.1.1,

$$P = [i_S - f(v_L)]v_L \qquad (3.1.3)$$

P has an extremum when $dP/dv_L = 0$:

$$i_S - f(v_L) - v_L \frac{df(v)}{dv}\bigg|_{v_L} = 0 \qquad (3.1.4)$$

J. Ladvánszky, *Theory of Power Matching*, SpringerBriefs in Electrical and
Computer Engineering, https://doi.org/10.1007/978-3-030-16631-1_3

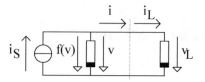

Fig. 3.1.1 Model of a loaded solar cell. Lower end of the nonlinear conductances is denoted by thick black line. In our example, i_S varies with light intensity, and $f(v)$ is an exponential diode characteristic

(3.1.4) holds when

$$g(v_L) = v_L \frac{df(v)}{dv}\bigg|_{v_L} \qquad (3.1.5)$$

3.2 Nonlinear, Resistive Maximum Power Theorem (Wyatt) and Its Reverse

Next, we formulate the nonlinear resistive power matching problem in general. In Theorem 3.2.1, sufficient conditions have been given. Consequences of the Theorem have been completed, and the necessary and sufficient version of the Theorem has been obtained (3.2.2). Concept of equivalence for nonlinear circuits has been introduced, and based on this, Norton and Thevenin equivalent of nonlinear, resistive multiports have been defined. In Theorem 3.2.3 we prove that generators having maximum power load possess exactly one Norton (Thevenin) equivalent that comprises independent sources and a passive, nonlinear multiport.

In Fig. 3.2.1, given the n-port N_G comprising independent sources and nonlinear resistors, it is to find the characteristics of the nonlinear resistive multiport N_L that dissipates maximum power at arbitrary voltages and currents of the independent voltage and current sources, respectively.

The solution has been given by the nonlinear resistive maximum power theorem [Wyatt83B]:

Theorem 3.2.1 We make the assumptions A.1, A.2, ...A.5:

A.1 Assume that the generator looks like that in Fig. 3.2.2., comprising the nonlinear, resistive n-port characterized by $\underline{I} = \underline{f}\left(\underline{V_G}\right)$ and independent current sources. Column matrices containing the port currents and voltages are denoted by $\underline{I}$ and $\underline{V_G}$, respectively:

$$\underline{I} = \begin{bmatrix} I_1 \\ I_2 \\ \vdots \\ I_n \end{bmatrix} \quad \underline{V_G} = \begin{bmatrix} V_{G1} \\ V_{G2} \\ \vdots \\ V_{Gn} \end{bmatrix} \qquad (3.2.1)$$

Fig. 3.2.1 Power matching problem of the nonlinear resistive generator N_G: It is to find the characteristics of the load N_L that dissipates maximum power at arbitrary element values of the independent sources

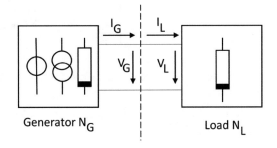

Generator N_G Load N_L

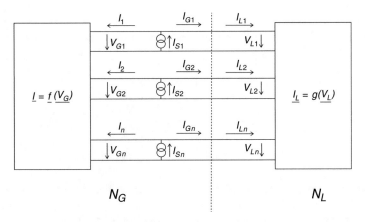

Fig. 3.2.2 The power matching problem for a generator of Norton structure

A.2 The generator conductance characteristic $\underline{f}(.)$ is continuously differentiable everywhere.

A.3 The function $\underline{h}(.)$ introduced as

$$\underline{h}\left(\underline{V_G}\right) = \underline{f}\left(\underline{V_G}\right) + \underline{\underline{J_f^T}}\left(\underline{V_G}\right)\underline{V_G} \tag{3.2.2}$$

is strictly monotonically increasing:

$$\left[\underline{h}\left(\underline{V_1}\right) - \underline{h}\left(\underline{V_2}\right)\right]^T \left(\underline{V_1} - \underline{V_2}\right) > 0 \tag{3.2.3}$$

for any $\underline{V_1}$ and $\underline{V_2}$. $\underline{\underline{J_f^T}}\left(\underline{V_G}\right)$ is the transpose of the Jacobian matrix of the function $\underline{f}$ at $\underline{V_G}$ whose entry in the pth row and rth column is:

$$\left[\underline{\underline{J}}_f^T\left(\underline{V}_G\right)\right]_{pr} = \frac{\partial f_r(\underline{V})}{\partial V_p}\Bigg|_{\underline{V}=\underline{V}_G} \tag{3.2.4}$$

A.4 The function $\underline{h}(.)$ is a surjection or onto mapping, that is, for every $\underline{h}$, there is a $\underline{V}$ so that $\underline{h} = \underline{h}(\underline{V})$ [Dieudonné69].

A.5 The characteristics of the load $\underline{I}_L = \underline{g}\left(\underline{V}_L\right)$ is given by

$$\underline{g}\left(\underline{V}_L\right) = \underline{\underline{J}}_f^T\left(\underline{V}_L\right)\underline{V}_L \tag{3.2.5}$$

where $\underline{I}_L$ and $\underline{V}_L$ stand for the column matrices of the load currents and voltages, respectively.

$$\underline{I}_L = \begin{bmatrix} I_{L1} \\ I_{L2} \\ \vdots \\ I_{Ln} \end{bmatrix} \qquad \underline{V}_L = \begin{bmatrix} V_{L1} \\ V_{L2} \\ \vdots \\ V_{Ln} \end{bmatrix} \tag{3.2.6}$$

When the assumptions A.1–A.5 hold, the circuit in Fig. 3.2.2 has the following properties P.1–P.3:

P.1 The circuit has at least one operating point for arbitrary $\underline{I}_S$.

P.2 There is only one operating point.

P.3 At this operating point, the generator obtains maximum power.

Proof The circuit in Fig. 3.2.2 can be characterized by the following equations:

$$\underline{I}_S - \underline{f}\left(\underline{V}_S\right) = \underline{g}\left(\underline{V}_L\right) \tag{3.2.7}$$

$$\underline{V}_S = \underline{V}_L \tag{3.2.8}$$

Substituting (3.2.5) into (3.2.7) and using the notation (3.2.2) yields

$$\underline{I}_S = \underline{h}\left(\underline{V}_G\right) \tag{3.2.9}$$

According to A.4, $\underline{h}(.)$ is a surjection; thus, for any $\underline{I}_S$ there is at least one operating point characterized by $\hat{\underline{V}}_G$ (P.1).

For the same source currents, existence of two different operating points $\hat{\underline{V}}_{G1} \neq \hat{\underline{V}}_{G2}$ contradicts to the strict monotonicity of $\underline{h}(.)$ (A.3); therefore only one operating point exists (P.2).

Power obtained by the generator is

$$P\left(\underline{V_G}\right) = \left[\underline{I_S} - \underline{f}\left(\underline{V_G}\right)\right]^T \underline{V_G} \qquad (3.2.10)$$

Its gradient with respect to $\underline{V_G}$ is

$$\frac{\partial P}{\partial \underline{V_G}} = \underline{I_S} - \underline{h}\left(\underline{V_G}\right) \qquad (3.2.11)$$

We can see from (3.2.9) that in the operating point $\underline{\hat{V}_G}$, the power gradient is zero:

$$\left.\frac{\partial P}{\partial \underline{V_G}}\right|_{\underline{V_G}=\underline{\hat{V}_G}} = 0 \qquad (3.2.12)$$

The operating point $\underline{\hat{V}_G}$ from (3.2.9) depends on $\underline{I_S}$. Choosing the load characteristic (3.2.5) provides that the gradient is zero for arbitrary $\underline{I_S}$ that is necessary condition for power maximum. We show that here it is also sufficient, so that in the operating point $\underline{\hat{V}_G}$, generator power has a global maximum for arbitrary $\underline{I_S}$:

$$P(\underline{\hat{V}_G}) - P\left(\underline{V_G}\right) > 0 \qquad (3.2.13)$$

for any $\underline{V_G} \neq \underline{\hat{V}_G}$. Transformation of the arguments yields

$$P(\underline{\hat{V}_G}) - P\left(\underline{V_G}\right) = P\left[\underline{\hat{V}_G} + \lambda(\underline{V_G} - \underline{\hat{V}_G})\right]_{\lambda=0} - P\left[\underline{\hat{V}_G} + \lambda(\underline{V_G} - \underline{\hat{V}_G})\right]_{\lambda=1} \qquad (3.2.14)$$

$$P(\underline{\hat{V}_G}) - P\left(\underline{V_G}\right) = -\int_0^1 \left\{\frac{d}{d\lambda}\left[\underline{\hat{V}_G} + \lambda(\underline{V_G} - \underline{\hat{V}_G})\right]\right\}d\lambda \qquad (3.2.15)$$

Within the argument, differentiation is performed using the chain rule:

$$P(\underline{\hat{V}_G}) - P\left(\underline{V_G}\right) = -\int_0^1 \left.\frac{dP}{d\underline{V_G}}\right|_{\underline{\hat{V}_G}+\lambda(\underline{V_G}-\underline{\hat{V}_G})}^T (\underline{V_G} - \underline{\hat{V}_G})d\lambda \qquad (3.2.16)$$

Using (3.2.12),

$$P(\hat{\underline{V}}_G) - P(\underline{V}_G) = -\int_0^1 \left[\left. \frac{dP}{dV_G} \right|^T_{\hat{\underline{V}}_G + \lambda(\underline{V}_G - \hat{\underline{V}}_G)} - \left. \frac{dP}{dV_G} \right|^T_{\hat{\underline{V}}_G} \right] (\underline{V}_G - \hat{\underline{V}}_G) d\lambda$$

$$(3.2.17)$$

Using (3.2.11) and identically rewriting the second term on the right:

$$P(\hat{\underline{V}}_G) - P(\underline{V}_G) = \int_0^1 \left[\underline{h}^T\left(\hat{\underline{V}}_G + \lambda(\underline{V}_G - \hat{\underline{V}}_G)\right) - \underline{h}^T(\hat{\underline{V}}_G) \right] \left[\hat{\underline{V}}_G + \lambda(\underline{V}_G - \hat{\underline{V}}_G) - \hat{\underline{V}}_G \right]$$

$$\times \frac{1}{\lambda} d\lambda =$$

$$= \int_0^1 \left[\underline{h}^T(\tilde{\underline{V}}_G) - \underline{h}^T(\hat{\underline{V}}_G) \right] \left[\tilde{\underline{V}}_G - \hat{\underline{V}}_G \right] \frac{1}{\lambda} d\lambda \qquad (3.2.18)$$

where the notation $\tilde{\underline{V}}_G = \hat{\underline{V}}_G + \lambda(\underline{V}_G - \hat{\underline{V}}_G)$ has been introduced. From strict monotonicity of $\underline{h}(.)$ defined by (3.2.3) follows that the integrand is always positive. The integral exists because $\underline{h}(.)$ is continuous by A.2. Therefore (3.2.3) and P.3 has been proved.

Next, we call optimum load, the load dissipating maximum power for arbitrary $\underline{I}_S$.

Note that for fixed $\underline{I}_S$, only one point of the characteristics of the optimum load is given. Thus, for fixed source currents, characteristics of the optimum load are not necessarily the same as in (3.2.5). Essence of Theorem 3.2.1 is that characteristics of the optimum load can also be given even if $\underline{I}_S$ is not fixed.

We start our own results with adding two other properties of the optimum load:
P.4 Load characteristics $\underline{g}(.)$is continuous.
P.5 Load current $\underline{I}_L$ and voltage $\underline{V}_L$ do not depend (cutting the interconnection to the generator in Fig. 3.2.2) on any of the variable parameters P_k of the generator:

$$\frac{\partial \underline{I}_L}{\partial P_k} = \underline{0} \quad \frac{\partial \underline{V}_L}{\partial P_k} = \underline{0} \qquad (3.2.19)$$

where $k = 1, 2,\ldots n_p$ and n_p is the number of parameters.

Theorem 3.2.2 (Nonlinear resistive maximum power theorem and its reverse): For the circuit in Fig. 3.2.1, A.1–A.5 hold if and only if P.1–P.5 hold.

Proof Sketch of the proof is given in Fig. 3.2.3. Firs three lines of necessity have been treated above.

Necessity	Sufficiency
A.1, A.5, A.4 → P.1	P.5, P.3→ A.1
A.3, P.1→ P.2	P.1, P.2, P.3 → A.3
P.1, A.3, A.5 → P.3	P.1 → A.4
A.2, A.5→ P.4	P.3, A.1, P.4 → A.5
A.5→ P.5	P.4, A.5→ A.2

Fig. 3.2.3 Sketch of the proof of Theorem 3.2.2. A.1-A5 and P.1-P5 denote the assumptions and properties in the text, respectively

Generator power is $P_G = I_G^T V_G$ where $\underline{I_G} = \underline{I_G}\left(\underline{V_G}, \underline{P}\right)$ where $\underline{P}$ is the column matrix of the variable parameters of the generator. Maximum power condition P.3 can be written as

$$\underline{I_G} + \frac{\partial \underline{I_G}^T}{\partial \underline{V_G}} \underline{V_G} = \underline{0} \tag{3.2.20}$$

Using in Fig. 3.2.1, relations $\underline{I_G} = \underline{I_L}$ and $\underline{V_G} = \underline{V_L}$ following from the interconnection between the generator and the load, differentiating both sides of (3.2.20) with respect to P_k, yield

$$\frac{\partial \underline{I_L}}{\partial P_k} + \frac{\partial \underline{I_G}^T}{\partial \underline{V_G}} \frac{\partial \underline{V_L}}{\partial P_k} + \frac{\partial^2 \underline{I_G}^T}{\partial P_k \partial \underline{V_G}} \underline{V_L} = \underline{0} \tag{3.2.21}$$

Comparison between (3.2.19) and (3.2.21) yields, for $\underline{V_G} \neq \underline{0}$:

$$\frac{\partial^2 \underline{I_G}^T}{\partial P_k \partial \underline{V_G}} \underline{V_L} = \underline{\underline{0}} \tag{3.2.22}$$

Accordingly, expression of the generator current must not contain a term depending on both $\underline{V_G}$ and P_k:

$$\underline{I_G}\left(\underline{V_G}, \underline{P}\right) = \underline{A}\left(\underline{V_G}\right) + \underline{B}\left(\underline{P}\right) \tag{3.2.23}$$

Equation (3.2.23) with substitutions $\underline{f}\left(\underline{V_G}\right) = -\underline{A}\left(\underline{V_G}\right)$ and $\underline{I_S} = \underline{B}\left(\underline{P}\right)$ gives the characteristics of the generator in Fig. 3.2.2. Thus, we proved that P.3 and P5 results in A.1.

We proved the first line of necessity in Fig. 3.2.3. Proof of the second and third lines is trivial. Later we return to the proof of lines 4 and 5.

In case of power matching of a solar cell, and many other applications that we investigate, power dissipated by the load is positive. According to the assumption A.5, load power can be expressed in terms of load voltage:

$$P_L = \underline{V_L}^T \underline{\underline{J_f^T}}\left(\underline{V_L}\right)\underline{V_L} \qquad (3.2.24)$$

It follows from (3.2.24) and $P > 0$ that the characteristics of the generator conductance is strictly monotonically increasing:

$$\left[\underline{f}\left(\underline{V_1}\right) - \underline{f}\left(\underline{V_2}\right)\right]^T \left(\underline{V_1} - \underline{V_2}\right) > 0 \qquad (3.2.25)$$

for any $\underline{V_1}$ and $\underline{V_2}$. Let $\underline{V_2} = 0$:

$$\left[\underline{f}\left(\underline{V}\right) - \underline{f}\left(\underline{0}\right)\right]^T \underline{V} > 0 \qquad (3.2.26)$$

Physical meaning of (3.2.26) is that the nonlinear conductance $\underline{f}\left(\underline{V}\right) - \underline{f}\left(\underline{0}\right)$ is passive.

Essence of the Theorem 3.2.2 is that if the generator N_G in Fig. 3.2.1 has an optimum load, then it can be replaced by a circuit shown on the left of Fig. 3.2.2 that is similar to the Norton equivalent of linear circuits. This is the motivation for the following definition.

Definition 3.2.1 Given two nonlinear resistive generators denoted by N_{G1} and N_{G2}, assume there are optimum loads for both, denoted by N_{L1} and N_{L2}. We define N_{G1} and N_{G2} equivalent if their optimum loads are identical: $N_{L1} \equiv N_{L2}$.

Theorem 3.2.3 If the n-port N_G in Fig. 3.2.1 has an optimum load, it is passive and P.1–P.5 hold, then N_G has one and only one Norton-equivalent, whose conductance is passive.

The equivalence concept defined here tends to the known equivalence concept for linear circuits. For linear n-ports, admissible signal pairs [Kuh67] are the same as those of their Norton equivalents. However, for nonlinear circuits, admissible signal pairs of the n-ports and that of their Norton equivalent are not the same in general.

Example 3.2.1 This example shows the special feature of the definition of equivalence above, given the generator in Fig. 3.2.4/a comprising an independent voltage source and two linear resistors.

We assume that $R_2 > 0$. R_1 is variable, V and R_2 are fixed. Load characteristics maximizing dissipated power for arbitrary R_1 is

Fig. 3.2.4 (**a**) The generator investigated in Example 3.2.1. (**b**) Its Norton equivalent. (**c**) Its Thevenin equivalent. Variable parameters are encircled

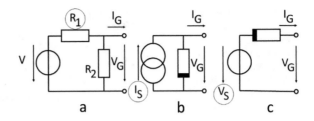

$$V_L = \frac{VR_2I_L}{V + 2R_2I_L} \quad I_L \neq \frac{-V}{2R_2} \quad V_L \neq \frac{V}{2} \tag{3.2.27}$$

From (3.2.5) we can determine the characteristics of the generator in Fig. 3.2.4b:

$$I_G = \frac{V}{2R_2}\ln\frac{V - 2V_G}{V} + I_S \quad V_G < \frac{V}{2} \text{ if } V > 0 \quad V_G > \frac{V}{2} \text{ if } V < 0 \tag{3.2.28}$$

Similarly, the characteristics of the generator in Fig. 3.2.4c can be obtained:

$$V_G = -\frac{V}{2}\ln\frac{V + 2R_2I_G}{V} + V_S \quad I_G < \frac{-V}{2R_2} \text{ if } V < 0 \quad I_G > \frac{-V}{2R_2} \text{ if } V > 0 \tag{3.2.29}$$

It should be emphasized that (3.2.27–3.2.29) are determined for arbitrary R_1; thus, R_1 is missing from (3.2.27–3.2.29).

In a comparison between (3.2.27) and (3.2.29) we can see that the set of admissible signal pairs of the original generator, its Norton and Thevenin equivalents are three different sets, whose common part is exactly the set of admissible signal pairs of the optimum load.

Example 3.2.2 Theorem 3.2.2 has been studied by computer simulation. A simple algorithm has been constructed that calculates the points of the optimum load characteristics from the generator characteristics $f(.)$. The algorithm is as follows. (3.2.11 and 3.2.12) for one-ports is:

$$0 = I_S - h(V) \tag{3.2.30}$$

In contrary with property P.2 we allow that (3.2.11) has multiple solutions: $\hat{V}_1$, $\hat{V}_2, \ldots \hat{V}_n$. Power corresponding to $\hat{V}_k$, $(k = 1, 2, \ldots n)$ is denoted by P_k

$$P_k = \left[I_S - f(\hat{V}_k)\right]\hat{V}_k \tag{3.2.31}$$

Assuming that the number of solutions n is finite, we can always find the voltage $\hat{V}_m$ $(1 \leq m \leq n)$ for that the power is maximum: $P_m = \max(P_1, P_2, \ldots P_n)$. Consequently, the point determined by $\hat{V}_m$ and $\hat{I}_m = I_S - f(\hat{V}_m)$ is surely on the optimum load characteristics. This algorithm has been applied for the example by [Wyatt83B]:

Fig. 3.2.5 (a) The function
$h(.)$ is not monotonic in
Example 3.2.2, (b) power
may have more maxima,
(c) characteristic of the
optimum load is not
continuous

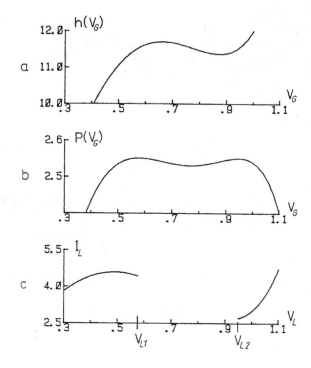

$$f(V_G) = 3V_G^5 + 9V_G^3 + 15V_G \qquad (3.2.32)$$

$$h(V_G) = 18V_G^5 - 36V_G^3 + 30V_G \qquad (3.2.33)$$

The function $h(V_G)$ in (3.2.33) has been plotted in Fig. 3.2.5a. We can see that $h(.)$ is not monotonic in contrary with A.3; thus, (3.2.30) may have more maxima. Thus, the generator power may have more extrema. In Fig. 3.2.5b, we achieved by properly selected I_S that the maximum values are identical, corresponding to two different values V_{L1} and V_{L2}.

Hurting the assumption A.3 resulted in that optimum load characteristics in Fig. 3.2.5c is not continuous. But even in that case, the optimum load exists because the curve in Fig. 3.2.5c can be computed. From this, we can conclude that the optimum load may exist in more general conditions than A.1 ... A.5. But a resistor with discontinuous characteristics cannot be realized; therefore continuity in P.4 and strict monotonicity in A.3 are in close connection with reality.

Of course, using a switch, we can realize discontinuous characteristics but that in Fig. 3.2.5c is not unicursal and that is the problem.

3.3 Reverse of the Power Maximizing Problem (Ladvánszky)

Up to this point, we investigated how to find optimum load for a given generator N_G.

Now we investigate the reverse problem. Given a nonlinear n-port N_L, it is to find N_G so that for all values of variable parameters in N_G, dissipated power by N_L is maximum. Realization example for variable parameters has been shown in Example 3.2.1.

The n-port N_G that meets the above conditions will be called as optimum generator. According to Theorem 3.2.3, N_G has a Norton equivalent whose inner conductance is passive. Characteristics of the passive inner conductance will be determined now from the characteristics of N_L.

In our calculations, results from vector analysis have been used. The following notations have been used: $\underline{a}$ and a_k denote an n component vector and its kth Descartes coordinate, respectively. Components of the nabla operator are the partial derivatives with respect to the independent variables: $\underline{\nabla} = \left(\frac{\partial}{\partial x_1}, \frac{\partial}{\partial x_2}, \ldots \frac{\partial}{\partial x_n} \right)$, if the independent variables have been denoted by x_1, $x_2, \ldots x_n$. Scalar and vector products have been defined by $\underline{a}\,\underline{b} = \sum_{k=1}^{n} a_k b_k$, $\underline{a} \times \underline{b} = (a_2 b_3 - a_3 b_2, a_3 b_1 - a_1 b_3, a_1 b_2 - a_2 b_1)$, respectively. The nabla operator and the scalar product have been defined for arbitrary n while the vector product for $n = 3$. Thus, first we calculate the characteristics of the generator for $n \leq 3$, then we extend them for arbitrary n.

Let $\underline{y}(x)$ and $\underline{z}(x)$ be three-dimensional, differentiable vector spaces. For the gradient of $\underline{y}\,\underline{z}$, the following identity holds [Korn75, p. 159]:

$$\underline{\nabla} \left(\underline{y}\,\underline{z} \right) = \left(\underline{y}\,\underline{\nabla} \right)\underline{z} + \left(\underline{z}\,\underline{\nabla} \right)\underline{y} + \underline{y} \times \left(\underline{\nabla} \times \underline{z} \right) + \underline{z} \times \left(\underline{\nabla} \times \underline{y} \right) \qquad (3.3.1)$$

Apply this identity for the gradient of power of the generator in Fig. 3.2.2, for $n \leq 3$.

$$P = \left[\underline{I}_S - \underline{f}(\underline{V}) \right]\underline{V} \qquad (3.3.2)$$

Substituting $\underline{x} = \underline{z} = \underline{V}$, $\underline{y} = \underline{I}_S - \underline{f}(\underline{V})$, yields for the condition of power maximum:

$$\underline{I}_S - \underline{f} - \left(\underline{V}\,\underline{\nabla} \right)\underline{f} - \underline{V} \times \left(\underline{\nabla} \times \underline{f} \right) = \underline{0} \qquad (3.3.3)$$

As $\underline{g}(\underline{V}) = \underline{I}_S - \underline{f}(\underline{V})$, load characteristic is deduced from (3.3.3):

$$\underline{g}(\underline{V}) = \left(\underline{V} \ \nabla \right)\underline{f} + \underline{V} \times \left(\nabla \times \underline{f} \right) \tag{3.3.4}$$

Substituting (3.3.4) into (3.3.3) yields

$$\underline{I}_S - \underline{f} - \underline{g} = \underline{0} \tag{3.3.5}$$

Applying the operator $\nabla \times$ for both sides of (3.3.5):

$$\nabla \times \underline{g} = - \nabla \times \underline{f} \tag{3.3.6}$$

Substituting (3.3.6) into (3.3.4), after rearrangement:

$$\left(\underline{V} \ \nabla \right)\underline{f} = \underline{g} + \underline{V} \times \left(\nabla \times \underline{g} \right) \tag{3.3.7}$$

(3.3.7) is a linear, first order, inhomogenous system of partial differential equations, whose general solution is known [Freud58]. A possible solution is the following:

$$\underline{F}(\underline{V}_0, t) = \int \underline{g} + \underline{V} \times \left(\nabla \times \underline{g} \right)\Big|_{\underline{V}=\underline{V}_0 e^t} dt \tag{3.3.8}$$

$$\underline{f}(\underline{V}) = \underline{F}(\underline{V}_0, t)\big|_{\underline{V}_0 = V e^{-t}} + \underline{C} \tag{3.3.9}$$

The constant $\underline{C}$ is selected so that $\underline{f}(\underline{0}) = \underline{0}$ is satisfied. According to Theorem 3.2.3, this is for the passivity of the generator conductance.

The calculation above is valid for $n \leq 3$. For two-ports, $\underline{V} \times \left(\nabla \times \underline{g} \right)$ is the following:

$$\underline{V} \times \left(\nabla \times \underline{g} \right) = \left[V_2\left(\frac{\partial g_2}{\partial V_1} - \frac{\partial g_1}{\partial V_2} \right), V_1\left(\frac{\partial g_1}{\partial V_2} - \frac{\partial g_2}{\partial V_1} \right) \right] \tag{3.3.10}$$

For one-ports, (3.3.8) and (3.3.9) are simplified into the following, known formula:

$$f(V_G) = \int\limits_{0}^{V_G} \frac{g(V)}{V} dV + C \tag{3.3.11}$$

Vector products in (3.3.7) are defined for $n \leq 3$. Therefore, to extend our calculation for greater n, we have to find the extension of (3.3.7) for greater n.

Let us make the gradient of the scalar field $(\underline{V}\,\underline{f})$ using the identity (3.3.1):

$$\nabla \left(\underline{V}\underline{f}\right) = \left(\underline{V}\ \nabla\right)\underline{f} + \underline{f} + \underline{V} \times \left(\nabla \times \underline{f}\right) \tag{3.3.12}$$

Rearranged as:

$$\underline{V} \times \left(\nabla \times \underline{f}\right) = \nabla \left(\underline{V}\underline{f}\right) - \underline{f} - \left(\underline{V}\ \nabla\right)\underline{f} \tag{3.3.13}$$

The expression at the right side is valid for arbitrary n. Thus (3.3.13) is considered in the following as the definition of $\underline{V} \times \left(\nabla \times \underline{f}\right)$ for arbitrary n. Its kth component is:

$$\underline{V} \times \left(\nabla \times \underline{f}\right)\Big|_{k} = \frac{\partial}{\partial V_k} \sum_{l} V_l f_l - f_k - \sum_{l} V_l \frac{\partial f_k}{\partial V_l} = \sum_{l} V_l \left(\frac{\partial f_l}{\partial V_k} - \frac{\partial f_k}{\partial V_l}\right) \tag{3.3.14}$$

Now (3.2.5) will be transformed to a form like (3.3.4), using (3.3.14):

$$g_k = \sum_{l} \frac{\partial f_l}{\partial V_k} V_l = \sum_{l} V_l \frac{\partial f_k}{\partial V_l} + \sum_{l} V_l \left(\frac{\partial f_l}{\partial V_k} - \frac{\partial f_k}{\partial V_l}\right) \tag{3.3.15}$$

For further step, (3.3.6) has to be generalized. This will be done using (3.2.5):

$$\frac{\partial g_l}{\partial V_k} - \frac{\partial g_k}{\partial V_l} = -\left(\frac{\partial f_l}{\partial V_k} - \frac{\partial f_k}{\partial V_l}\right) \tag{3.3.16}$$

Based on the last two equations, (3.3.7) can be generalized as

$$\sum_{l} V_l \frac{\partial f_k}{\partial V_l} = g_k + \sum_{l} V_l \left(\frac{\partial g_l}{\partial V_k} - \frac{\partial g_k}{\partial V_l}\right) \tag{3.3.17}$$

from which $\underline{f}(\underline{V})$ can be determined as shown above.

Now we can continue the proof of Theorem 3.2.2. First, based on (3.3.16), we complete the properties of the optimum load:

P.4′ The load characteristics $g(.)$ are continuous. For more than one ports, the differences in (3.3.16) can be expressed and they are continuous.

For one ports, it follows from P.4 that $g(V_L)/V_L$ is continuous everywhere except the origin thus integrable. Therefore, the characteristics in (3.3.11) exist, and if we chose that as the characteristics of the optimum generator, A.2 and A.5 are fulfilled.

For more than one ports, the expression at the right of (3.3.17) is continuous as a consequence of P.4′; thus, the integral in (3.3.8) exists. Choosing the optimum generator characteristics as in (3.3.8 and 3.3.9), A.2 and A.5 are satisfied.

Thus, we proved in Fig. 3.2.3, the fourth and fifth lines of necessity; thus, the proof is completed.

3.4 Properties of Corresponding Source-Load Pairs (Ladvánszky)

An arbitrary resistive n-port can be decomposed as an interconnection of reciprocal and solenoidal n-ports [Wyatt78]. The conductance $\underline{I} = \underline{g}(\underline{V})$ is reciprocal if and only if its differential conductance matrix is symmetrical:

$$G_{kl}(\underline{V}) = \frac{\partial g_k}{\partial V_l} = G_{lk}(\underline{V}) \tag{3.4.1}$$

The conductance $\underline{I}=\underline{g}(\underline{V})$ is solenoidal if its characteristics are divergence-free:

$$\sum_l \frac{\partial g_l}{\partial V_l} = 0 \tag{3.4.2}$$

Theorem 3.4.1 The generator conductance is reciprocal if and only if the characteristics of the optimum load is.

Proof The statement is the consequence of (3.3.16).

Is there a relation between the solenoidal properties of the generator and the optimum load? The answer is negative:

$$\nabla \underline{g} = \sum_k \frac{\partial g_k}{\partial V_k} = \sum_k \frac{\partial}{\partial V_k} \sum_l V_l \frac{\partial f_l}{\partial V_k} = \sum_k \frac{\partial f_k}{\partial V_k} + \sum_{k,l} V_l \frac{\partial^2 f_l}{\partial V_k^2}$$

$$= \sum_k \frac{\partial f_k}{\partial V_k} + \sum_l V_l \frac{\partial^2 f_l}{\partial V_l^2} + \sum_l V_l \left(\sum_{k \neq l} \frac{\partial^2 f_l}{\partial V_k^2} \right) =$$

$$= \sum_k \frac{\partial f_k}{\partial V_k} + \sum_l V_l \frac{\partial^2 f_l}{\partial V_l^2} + \sum_l V_l \left(\sum_{k \neq l} \frac{\partial^2 f_k}{\partial V_l \partial V_k} \right) - \sum_l V_l \left(\sum_{k \neq l} \frac{\partial^2 f_k}{\partial V_l \partial V_k} \right)$$

$$+ \sum_l V_l \left(\sum_{k \neq l} \frac{\partial^2 f_l}{\partial V_k^2} \right) =$$

$$= \sum_k \frac{\partial f_k}{\partial V_k} + \sum_l V_l \frac{\partial}{\partial V_l} \left(\sum_k \frac{\partial f_k}{\partial V_k} \right) - \sum_l V_l \sum_k \frac{\partial}{\partial V_k} \left(\frac{\partial f_k}{\partial V_l} - \frac{\partial f_l}{\partial V_k} \right) \quad (3.4.3)$$

For a solenoidal generator ($\nabla \underline{f} = 0$) the optimum load is not surely solenoidal because at the last row of (3.4.3), the first two terms are zero, the last term may differ from zero. Another conclusion from (3.4.3) is that if the generator is solenoidal and reciprocal, then the optimum load is solenoidal and reciprocal.

Rearrangement of (3.3.16) yields

$$\frac{\partial (f_l + g_l)}{\partial V_k} = \frac{\partial (f_k + g_k)}{\partial V_l} \quad (3.4.4)$$

Physical meaning of (3.4.4) is:

Theorem 3.4.2 Parallel interconnection of the generator and its optimum load always results in a reciprocal n-port.

This is surprising because the differential conductance matrix depends on the port voltages. The same n-port can be reciprocal or nonreciprocal depending on the voltage. The property according to Theorem 3.4.2 holds for arbitrary voltages and arbitrary generator having optimum load, however.

3.5   Direct Measurement of the Thermal Voltage of Semiconductor Diodes (Ladvánszky)

Characteristics of a pn junction diode for small current and low frequency is

$$i_D = I_0 \left(e^{\frac{v_D}{\lambda V_T}} - 1 \right) \quad (3.5.1)$$

where i_D, v_D, I_0, and λV_T are the diode current, diode voltage, saturation current, and modified thermal voltage, respectively. At room temperature, $V_T = 26 \ mV$ for an abrupt pn junction. λ is the modification factor, without modification, $\lambda = 1$. The shape of the characteristics is determined by I_0 and λV_T. Both can be determined using at least two measured i_D, v_D pairs. However, in the following we show that λV_T can be directly measured.

Fig. 3.5.1 Diode
measurement circuit

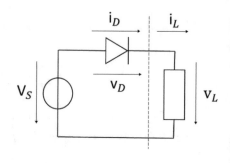

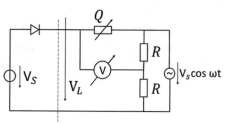

Fig. 3.5.2 The diode, the potentiometer Q, and the two resistors R constitute a Wheatstone bridge.
If the bridge is balanced, then the DC component of the load voltage V_L is equal to the modified
thermal voltage of the diode λV_T

Look at Fig. 3.5.1 showing a nonlinear generator built up with a voltage source
and a diode, loaded by a linear resistor. Measurement principle is based on three
statements. The first one is a consequence of the A.5 assumption of the maximum
power theorem.

Rearranging (3.2.5) written for Thevenin equivalents, we obtain

$$\frac{v_L}{i_L} = f'(i_G) \tag{3.5.2}$$

Statement 1 If the large signal resistance of the load is equal to the differential
resistance of the generator, then the load dissipates maximum power.

It can be controlled that in the circuit of Fig. 3.5.1, for $v_L > 0$, other assumptions of
Theorem 3.2.1 are also fulfilled.

Statement 2 In Fig. 3.5.1, the load resistor dissipates maximum power when the
voltage across the resistor is equal to the modified thermal voltage λV_T of the diode.

Statement 3 (obvious) Large signal and differential resistances of a linear resistor
are identical.

For application of (3.5.2), in the circuit of Fig. 3.5.2, differential resistance of the
diode must appear; therefore the diode has been excited by a sinusoidal of small
amplitude:

Fig. 3.5.3 Block diagram
of the realized measurement
circuit. At the first
measurement, $V_l = V_s/2$ is
adjusted by V_R. No further
adjustment is necessary

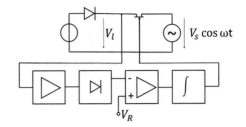

$$v_S = V_s \cos \omega t \qquad \frac{V_s}{2\lambda V_T} \ll 1 \qquad (3.5.3)$$

Because the excitation amplitude is small, it is enough to consider the DC and the first harmonic of the load voltage:

$$v_L = V_L + V_l \cos \omega t \qquad (3.5.4)$$

In Fig. 3.5.2, the load was replaced by the circuit shown at the right of the dashed line. The sinusoidal voltage source realizes a short circuit at DC; therefore the DC model of the circuit is identical to that in Fig. 3.5.1. The DC voltage source at the left of Fig. 3.5.2 can be replaced by a short circuit at AC frequencies. Thus, the diode, the potentiometer Q, and the two resistors R constitute a Wheatstone bridge whose balance is indicated by a sensitive AC voltage meter. If the bridge is balanced, then the resistance of the potentiometer Q is equal to the differential resistance of the diode, that is, maximum power is dissipated by the load. In this case $V_L = \lambda V_T$.

Channel resistance of FETs is linear for small channel voltages; it can be controlled by the gate-source voltage. Thus, balancing of the bridge can be automated according to Fig. 3.5.3.

If the voltage amplitude V_s is kept constant, the two resistors R in Fig. 3.5.2 can be left out. AC voltage across the measured diode, after amplification and rectification, is compared to V_R. The difference between V_s and V_R is integrated and controls the gate to source voltage of the FET. When $V_R = V_s$, the channel resistance of the FET is equal to the differential resistance of the diode. In this case, the DC voltage of the channel is exactly equal to the modified thermal voltage of the diode to be measured.

Limitations of our measurement principle are the series resistance and the reactances of the measured diode. These secondary effects have been decreased by selecting a proper DC current of the measured diode and the measurement frequency. Analysis is shown in Fig. 3.5.4. FET type is 2N7000, and the opamps can be parts of ML324 quad.

Measurement frequency is 15 kHz and AC voltage amplitude is 5 mV. Diode current was adjusted to 1 mA. Several diodes have been measured. Modified thermal voltage has been given for the analyzed diode model and the value has been compared to the output of the circuit at the test point 1 (Table 3.5.1). Agreement is acceptable.

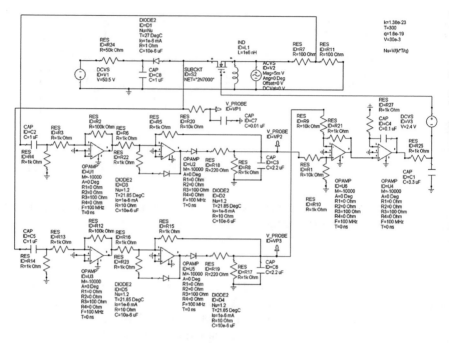

Fig. 3.5.4 The measurement circuit analyzed in AWR

Table 3.5.1 Comparison of the values of modified thermal voltage given for the analyzed diode model, and output at test point 1 of Fig. 3.5.4

Serial number of the diode	Given λV_T, mV	λV_T, analyzed in the next Figure, mV
1	26	27.44
2	27	27.95
3	28	28.43
4	29	28.94
5	30	29.4

Note that satisfying Fig. 3.5.2 is applicable for the measurement of the differential resistance of an arbitrary nonlinear resistor.

3.6   Summary

In this chapter, power matching of nonlinear, resistive generators has been investigated. New definition has been provided for equivalence of active, nonlinear, resistive multiports that is a natural extension of equivalence of linear circuits. We have proved that the interconnection between Norton and Thevenin equivalents of nonlinear generators and their load dissipating maximum power is close. Two examples have been obtained. In the first one, equivalences of linear and nonlinear circuits have been compared. From the second example it became obvious that the maximum power theorem is valid for more general conditions, but in that case realizability problems may occur. Naturally, all statements valid for Norton equivalent is also valid for Thevenin equivalent and vice versa.

Note that our definition given for nonlinear n-ports is also valid for passive circuits. In this case, the definition must be applied for the n-port extended by independent sources. Further investigation is necessary for the extension of such multiports that do not have maximum power load.

We also obtained the solution of the reverse of the power maximization problem, in that optimum generator has to be found for a given load. From the characteristic of the given load, we determined the characteristic of the inner conductance of the Norton equivalent of the optimum generator. Circuit theoretical properties for corresponding generator-load pairs have been given. We proved that interconnection of a generator and its optimum load is always reciprocal.

A consequence of the maximum power theorem has been applied for direct measurement of the modified thermal voltage of junction diodes.

References

[Dieudonné69]   Dieudonné, *Foundations of Modern Analysis* (Academic Press, New York, 1969)

[Freud58]   Freud, in *Parciális differenciálenyenletek (Partial differential equations, in Hungarian)*, ed. by F. Fazekas. Mathematical excercises, B. VIII (Tankönyvkiadó, Budapest, 1958)

[Korn75]   K. Korn, *Matematikai kézikönyv műszakiaknak (Mathematical Handbook for Technicians, in Hungarian)* (Műszaki Könyvkiadó, Piliscsév, 1975)

[Kuh67]   E.S. Kuh, R.A. Rohrer, *Linear Active Circuits* (Holden-Day, San Francisco, 1967)

[Wyatt78]   Wyatt-Chua-Oster, Nonlinear n-port decomposition via the Laplace operator. IEEE Trans. CAS. 25, 741–754 (1978)

[Wyatt83B]   J.L. Wyatt, L.O. Chua, Nonlinear resistive maximum power theorem, with solar cell application. IEEE Trans. CAS 30, 824–828 (1983)

Chapter 4
Linear Multiports, Competitive Power Matching (Lin)

4.1 Introduction

Problem statement is given in Fig. 4.1.1

The problem is formally the same as the so-called duopoly problem in economics [Lin85]. The related problem is the cooperative power matching when the loads are adjusted so that the total dissipated power is maximum [Lin72]. Cooperative total power is always greater than or equal to the competitive one.

4.2 The Original Problem and Its Solution

Equations for Fig. 4.1.1 are:

$$V_1 = E_1 + z_{11}I_1 + z_{12}I_2 \tag{4.2.1}$$

$$V_2 = E_2 + z_{21}I_1 + z_{22}I_2 \tag{4.2.2}$$

$$V_1 = -R_{L1}I_1 \tag{4.2.3}$$

$$V_2 = -R_{L2}I_2 \tag{4.2.4}$$

and all source impedances are real. Conditions: Load powers separately should be maximized so that the load resistors are adjusted consecutively.

It is known that the load power is maximum when the load resistor is adjusted to the same value seen from the source at the same port. In the kth step, $k \geq 1$:

$$R_{L1}(k) = z_{11} - \frac{z_{21}z_{12}}{z_{22} + R_{L2}(k-1)} \tag{4.2.5}$$

© The Author(s), under exclusive licence to Springer Nature Switzerland AG 2019
J. Ladvánszky, *Theory of Power Matching*, SpringerBriefs in Electrical and
Computer Engineering, https://doi.org/10.1007/978-3-030-16631-1_4

Fig. 4.1.1 Competitive
power matching: Problem
statement. The two load
resistors are adjusted
alternately so that each
separately dissipates
maximum power

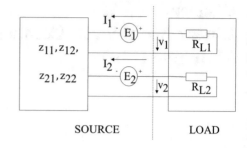

$$R_{L2}(k) = z_{22} - \frac{z_{21}z_{12}}{z_{11} + R_{L1}(k)} \tag{4.2.6}$$

In the stationary point:

$$R_{L1} = z_{11} - \frac{z_{21}z_{12}}{z_{22} + R_{L2}} \tag{4.2.7}$$

$$R_{L2} = z_{22} - \frac{z_{21}z_{12}}{z_{11} + R_{L1}} \tag{4.2.8}$$

That is, the load resistances are the image resistances of the source.

4.3   Generalization for Non-resistive Two-Ports

The problem statement is the same except that the source impedances are complex,
provided that the impedance matrix is positive definite, and the load impedances are
also complex. By definition of the image impedances, the result is the same: At the
stationary point, the load impedances are equal to the image impedances
[Wikipedia18] of the source:

$$Z_{L1}^* = z_{11} - \frac{z_{21}z_{12}}{z_{22} + Z_{L2}} \tag{4.3.1}$$

$$Z_{L2}^* = z_{22} - \frac{z_{21}z_{12}}{z_{11} + Z_{L1}} \tag{4.3.2}$$

where the star denotes complex conjugate. For the existence of the image impedances, the source should be unconditionally stable.

4.4 Application Example: Nash Equilibrium and Related Problems

This is a group of interesting problems. Obviously, the competitive power matching is related to the n participant generalization of the duopoly problem in economics [Henderson71] while the cooperative power matching is related to the Nash equilibrium. This is also a warning how to determine the Nash equilibrium in terms of circuit theory.

We prove that the n-port competitive power matching results in less or equal total load power than the cooperative one. It is worth cooperating!

The proof is very simple. In cooperative case, we extract from the source the maximum power obtainable. That means the competitive total power should be less or equal.

The proof has another important consequence. The maximum obtainable power can surely be achieved by cooperation while in competitive case, only in some special cases.

4.5 Summary

Competitive power matching has been studied. The loads are single resistors and adjusted consecutively to maximize its own dissipated power. The solutions are the image parameters of the source. The solution has been extended for sources with memory. We proposed that the circuit analogy for economic problems can be used for determining the Nash equilibrium numerically.

References

[Lin85]   P.M. Lin, Competitive power extraction from linear n-ports. IEEE Trans. Circuits Syst. **CAS-32**(2), 185–191 (1985)
[Lin72]   P.M. Lin, Determination of available power from resistive multiports. IEEE Trans. Circuit Theory **CT-19**, 385–386 (1972)
[Wikipedia18] Wikipedia, Image parameters
[Henderson71] J.M. Henderson, R.E. Grandt, *Microeconomic Theory: A Mathematical Approach* (McGraw-Hill, New York, 1971)

Chapter 5
The Scattering Matrix (Belevitch Approach), with Application to Broadband Matching

5.1 Introduction of the Scattering Matrix

In basic circuit theory, the scattering matrix is introduced as the measure of deviation from power matched case [Belevitch48, Kuh67].

The working hypothesis is illustrated in Fig. 5.1.1.

We imagine that the port quantities $\mathbf{V}$ and $\mathbf{I}$ in the arbitrarily loaded case can be formulated as the superposition of incident and reflected port quantities. The incident ones are when the loading is optimal, and parts of them are reflected:

$$\mathbf{V}_r = \mathbf{V} - \mathbf{V}_i \tag{5.1.1}$$

$$\mathbf{I}_r = -(\mathbf{I} - \mathbf{I}_i) \tag{5.1.2}$$

where

$$\mathbf{I}_i = \frac{1}{2}\,\mathbf{r}^{-\frac{1}{2}}\mathbf{V}_S \tag{5.1.3}$$

$$\mathbf{V}_i = \frac{1}{2}\,\mathbf{z}^*\mathbf{r}^{-\frac{1}{2}}\mathbf{V}_S \tag{5.1.4}$$

describing the optimal matching, while $\mathbf{V}$ and $\mathbf{I}$ are the voltages and currents in the arbitrarily loaded case.

The voltage and current based scattering matrices of the load in the left of Fig. 5.1.1 are defined as follows:

$$\mathbf{V}_r = \mathbf{S}^V\mathbf{V}_i \tag{5.1.5}$$

$$\mathbf{I}_r = \mathbf{S}^I\mathbf{I}_i \tag{5.1.6}$$

J. Ladvánszky, *Theory of Power Matching*, SpringerBriefs in Electrical and Computer Engineering, https://doi.org/10.1007/978-3-030-16631-1_5

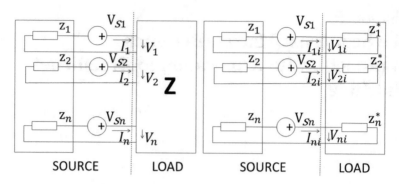

Fig. 5.1.1 Left: Arbitrary loading of n one-port sources. Right: optimal loading of n one-port sources

$\mathbf{S}^V$ and $\mathbf{S}^I$ can be expressed by the load impedance and admittance matrices as

$$\mathbf{S}^I = (\mathbf{Z} + \mathbf{z})^{-1}(\mathbf{Z} - \mathbf{z}^*) \tag{5.1.7}$$

$$\mathbf{S}^V = -(\mathbf{Y} + \mathbf{y})^{-1}(\mathbf{Y} - \mathbf{y}^*) \tag{5.1.8}$$

The normalized scattering matrix $\mathbf{S}$ can be defined using power waves:

$$\mathbf{a} = \mathbf{r}^{\frac{1}{2}}\mathbf{I}_i = \frac{1}{2}\,\mathbf{r}^{-\frac{1}{2}}(\mathbf{V} + \mathbf{z}\mathbf{I}) \tag{5.1.9}$$

$$\mathbf{b} = \mathbf{r}^{\frac{1}{2}}\mathbf{I}_r = \frac{1}{2}\,\mathbf{r}^{-\frac{1}{2}}(\mathbf{V} - \mathbf{z}^*\mathbf{I}) \tag{5.1.10}$$

Then:

$$\mathbf{b} = \mathbf{S}\mathbf{a} \tag{5.1.11}$$

$$\mathbf{V} = \mathbf{z}^*\mathbf{r}^{-\frac{1}{2}}\mathbf{a} + \mathbf{z}\mathbf{r}^{-\frac{1}{2}}\mathbf{b} \tag{5.1.12}$$

$$\mathbf{I} = \mathbf{r}^{-\frac{1}{2}}\mathbf{a} - \mathbf{r}^{-\frac{1}{2}}\mathbf{b} \tag{5.1.13}$$

For sinusoidal excitations, the power dissipated by the load is:

$$P = \mathrm{Re}(\mathbf{I}^{*T}\mathbf{V}) = (\mathbf{a}^{*T}\mathbf{a} - \mathbf{b}^{*T}\mathbf{b}) \tag{5.1.14}$$

Therefore, the circuit is lossless if and only if

$$\mathbf{S}^{*T}\mathbf{S} = \mathbf{1}_n \tag{5.1.15}$$

That is valid on the $j\omega$ axis.

5.2   Extension to the Whole Complex Frequency Plane (Youla)

Based on reasonable assumptions, the concept of the scattering matrix has been extended from the real frequency axis to the whole complex frequency plane. This step is necessary as a preparation for broadband matching.

Features of the extended scattering matrix have been proposed first. Then a possible way how to construct them has been described. Finally, a theorem concerning the realizability of the scattering matrix has been obtained.

1. Features [Youla64]

    The goal is to provide such an extension that preserves the values on the $j\omega$ axis and exhibits reasonable expectations.

    (a) $\mathbf{S}(p)$ is rational.
    (b) $\mathbf{S}(p)$ is analytic in $\mathrm{Re}(p) \geq 0$.
    (c) $\mathbf{1}_n - \mathbf{S}^{*T}(p)\mathbf{S}(p)$ is the matrix of a non-negative Hermitian quadratic form for all p in $\mathrm{Re}(p) \geq 0$.
    (d) For lossless n-ports, $\mathbf{S}_*\mathbf{S} = \mathbf{1}_n$, that is, $\mathbf{S}$ is paraconjugate unitary.
    (e) For a reciprocal n-port, S is symmetric: $\mathbf{S}^T = \mathbf{S}$.
    (f) The transducer power gain is $G_{rk}(\omega^2) = |s_{rk}(j\omega)|^2$, $r \neq k$.
    (g) If the z's are positive-real, $\mathbf{S}(p)$ is real for real p; and the paraconjugate unitary condition in (d) goes into the para-unitary condition $\mathbf{S}^T\mathbf{S} = \mathbf{1}_n$.

2. Construction

    Let $\mathbf{A}$ be an arbitrary rational matrix. Then we define

$$\mathbf{A}_*(p) = \mathbf{A}^{*T}(-p^*) \tag{5.2.1}$$

    If $\mathbf{A}$ is analytic in $\mathrm{Re}(p) > 0$, it is called regular.
    The regular all-pass function $\eta(p)$ has been introduced:

$$\eta(p) = \prod_{k=1}^{r} \frac{p - p_k}{p + p_k{}^*} \tag{5.2.2}$$

    Given the rational non-Foster functions $z_k(p)$ with positive the Hermitian part

$$r_k(p) = \frac{z_k(p) + z_{k*}(p)}{2} \tag{5.2.3}$$

    Then a factorization exists for Hurwitz and anti-Hurwitz rational functions:

$$r_k(p) = h_{k*}(p)h_k(p) \tag{5.2.4}$$

where $h_k(p) = \eta(p)\hat{h}_k(p)$ and $\hat{h}_k(p)$ does not have zeros and poles in $\mathrm{Re}(p) > 0$.

The incident and reflected waves are as follows:

$$2h_k(p)a_k(p) = V_k + z_k(p)I_k \tag{5.2.5}$$

$$2h_{k*}(p)b_k(p) = V_k - z_{k*}(p)I_k \tag{5.2.6}$$

The extension of the scattering matrix $S(p)$ has been defined as

$$\mathbf{b}(p) = \mathbf{S}(p)\mathbf{a}(p) \tag{5.2.7}$$

Proof that (5.2.7) satisfies the requirements 1–7 above has been found in [Youla64].

3. A realizability theorem

   This is the basis of the analytic broadband matching.

   The $n \times n$ real rational matrix $S(p)$ is the scattering matrix of a lumped, passive, reciprocal n-port if and only if:

   (a) $S(p)$ is analytic in $\mathrm{Re}(p) \geq 0$.
   (b) $\mathbf{1}_n - \mathbf{S}^{*T}(p)\mathbf{S}(p)$ is positive definite on the $j\omega$ axis.
   (c) The matrix

   $$\mathbf{Y}_A(p) = \frac{1}{2} \mathbf{H}^{-1}(p)(\mathbf{D}(p) - \mathbf{S}(p))\mathbf{H}^{-1}(p) \tag{5.2.8}$$

   is analytic in $\mathrm{Re}(p) > 0$, where:

   $$\mathbf{H}(p) = \mathrm{diag}(h_1(p), h_2(p), \ldots h_n(p)) \tag{5.2.9}$$

   $$\mathbf{D}(p) = \mathbf{H}(p)\mathbf{H}_*^{-1}(p) \tag{5.2.10}$$

4. $S^T(p) = S(p)$.

   The proof is found in [Youla64].

5.3 Broadband Matching, Analytic Approach (Youla, Kuh-Rohrer, Helton)

Broadband matching [Youla64a, Kuh67] is the art of generating intentional minimum mismatch to achieve the desired bandwidth. Another version where the open left half of the complex frequency plane has been mapped into the unit circle has been found in [Helton81].

The generator impedance is a positive real constant, and the load impedance is given analytically.

Please consider Fig. 5.3.1.

Fig. 5.3.1 Problem
statement for broadband
matching

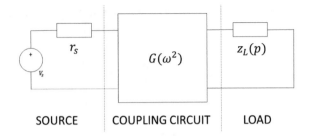

SOURCE   COUPLING CIRCUIT   LOAD

The goal is to realize the given $G(\omega^2)$ transducer power gain from the source to the load. $G(\omega^2) > 0$ is a real rational function of ω^2, the coupling circuit is assumed lossless. r_S is positive, $z_L(p)$ is positive real.

1. Specify $G(\omega^2)$ and construct the analytic continuation $G(-p^2)$ replacing ω^2 by $-p^2$ in the formula for $G(\omega^2)$.
2. Factorize $G(-p^2)$ as follows:

$$G(-p^2) = 1 - \Gamma(p)\Gamma(-p) \qquad (5.3.1)$$

where $\Gamma(p)$ is a bounded real rational reflection coefficient.
3. Realize $\Gamma(p)$ as the input reflection coefficient of a lossless two-port terminated by $z_L(p)$, according to the realizability theorem of Sect. 5.2.

The details are found in [Youla64].

An important point of realization is bandwidth. A typical specification for $G(\omega^2)$ is that it approximates a constant over a frequency band and zero elsewhere. The point is that the constant gain cannot be arbitrarily large. There is a maximum, and above that value, the problem cannot be solved using positive circuit elements in the lossless coupling circuit.

Note that fulfilling the conditions in [Youla64] or [Kuh67] can be replaced by investigation of the positivity of the realizing circuit elements.

5.4   Real Frequency Approach (Carlin)

Here the load admittance is given at separate frequency points ω_k and the source admittance is a positive real constant [Carlin77]. With the notations of the original paper, the transducer gain is:

$$T(\omega^2) = \frac{4R_L(\omega)R_q(\omega)}{|Z_L(\omega) + Z_q(\omega)|^2} \qquad (5.4.1)$$

where $Z_L(\omega) = R_L(\omega) + jX_L(\omega)$ is the given load impedance, $Z_q(\omega) = R_q(\omega) + jX_q(\omega)$ is the impedance of the coupling circuit seen from the load, to be computed. The transducer gain is prescribed at ω_k. We seek $Z_q(\omega)$ in the following (piecewise linear) form:

$$R_q(\omega) = r_0 + \sum_{k=1}^{n} a_k(\omega) r_k \qquad (5.4.2)$$

$$a_k(\omega) = \begin{cases} 1 & if \quad \omega_k \le \omega \\ \dfrac{\omega - \omega_{k-1}}{\omega_k - \omega_{k-1}} & if \quad \omega_{k-1} \le \omega \le \omega_k \\ 0 & if \quad \omega \le \omega_{k-1} \end{cases} \qquad (5.4.3)$$

$$X_q(\omega) = \sum_{k=1}^{n} b_k(\omega) r_k \qquad (5.4.4)$$

$$b_k(\omega) = \frac{1}{(\omega_k - \omega_{k-1})\pi} \int_{\omega_{k-1}}^{\omega_k} \ln\left|\frac{y + \omega}{y - \omega}\right| dy \qquad (5.4.5)$$

and we optimize for r_0 and r_k using (5.4.1–5.4.5). The final task is to synthetize $Z_q(\omega)$.

Intricates of the analytic broadband matching are bypassed. Details are found in [Carlin77].

5.5   Double Matching (Chien)

The linear power matching problem when both the generator and the load admittances are complex will be considered [Chien74].

Steps are the same as in Sect. 5.3. The difference is that now the most general form of a para-unitary matrix should be exploited. Later, this was simplified by Prof. A. Fettweis.

This solution is significant because other versions of broadband matching in Sects. 5.3 and 5.4, cannot be applied for interstage matching. A replacement of double matching can be provided by application of matching from complex to real and then from real to a different complex. However, this replacement results in unnecessarily complicated coupling circuits.

5.6   Summary

Introduction of the scattering matrix and its most important application in circuit design is treated: Broadband matching. Not all variants of broadband matching are included: Analytic, real frequency and double matching only. In addition to that, we suggest reading available papers about the parametric approach by Fettweis and Yarman, the real frequency version of double matching by Carlin, and soon, our approach based on image parameters. It has been ready for about 30 years but nobody else tried it.

References

[Belevitch48]   V. Belevitch, Transmission losses in 2n terminal networks. J. Appl. Phys. **19**(7), 636–638 (1948)

[Chien74]   T.-M. Chien, A theory of broadband matching of a frequency-dependent generator and Load-Part I. J. Franklin Inst., 181–199 (1974). https://doi.org/10.1016/0016-0032(74)90113-6

[Kuh67]   E.S. Kuh, R.A. Rohrer, *Theory of Linear Active Networks* (Holden-Day, Inc., San Francisco, 1967)

[Youla64]   D.C. Youla, An extension of the concept of scattering matrix. IEEE Trans. Circuit Theory **11**, 310–312 (1964)

[Youla64a]   D.C. Youla, A new theory of broadband matching. IEEE Trans. Circuit Theory **11**, 30–50 (1964)

[Helton81]   J.W. Helton, Broadbanding: Gain equalization directly from data. IEEE Trans. Circuits Syst. **CAS-28**(12), 1125–1137 (1981)

[Carlin77]   H.J. Carlin, A new approach to gain-bandwidth problems. IEEE Trans. Circuits Syst. **CAS-24**(4), 170–175 (1977)

Chapter 6
Foundation Concepts Based on Power Matching (Youla, Castriota, Carlin)

6.1 Existence of the Scattering Matrix

Sketch of the proof is that admittance matrix of an augmented n-port always exists, and the scattering matrix is expressed by the admittance matrix of the augmented n-port in the way that is always valid.

An augmented n-port is shown in Fig. 6.1.1.

Circuit equations for Fig. 6.1.1 are the following:

$$\mathbf{i}_A = \mathbf{i} \tag{6.1.1}$$

$$\mathbf{v}_A = \mathbf{v} + \mathbf{r}\mathbf{i} \tag{6.1.2}$$

$$\mathbf{Y}_A \mathbf{v}_A = \mathbf{i}_A \tag{6.1.3}$$

$$\mathbf{a} = \frac{1}{2} \mathbf{r}^{-\frac{1}{2}} (\mathbf{v} + \mathbf{z}\mathbf{i}) \tag{6.1.4}$$

$$\mathbf{b} = \frac{1}{2} \mathbf{r}^{-\frac{1}{2}} (\mathbf{v} - \mathbf{z}^* \mathbf{v}) \tag{6.1.5}$$

$$\mathbf{b} = \mathbf{S}\mathbf{a} \tag{6.1.6}$$

(6.1.1–6.1.6) are valid on the $j\omega$ axis. Notations r and z are diagonal matrices. The result is:

$$\frac{1}{2} \mathbf{r}^{-\frac{1}{2}} (\mathbf{v} - \mathbf{z}^* \mathbf{v}) = \mathbf{S} \frac{1}{2} \mathbf{r}^{-\frac{1}{2}} (\mathbf{v} + \mathbf{z}\mathbf{v})$$

J. Ladvánszky, *Theory of Power Matching*, SpringerBriefs in Electrical and Computer Engineering, https://doi.org/10.1007/978-3-030-16631-1_6

Fig. 6.1.1 Augmented
n-port

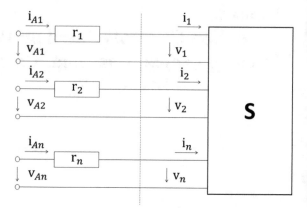

$$\mathbf{r}^{-\frac{1}{2}}(\mathbf{1}_n - (\mathbf{r} + \mathbf{z}^*)\mathbf{Y}_A) = \mathbf{S}\mathbf{r}^{-\frac{1}{2}}(\mathbf{1}_n - (\mathbf{r} - \mathbf{z})\mathbf{Y}_A)$$

$$\mathbf{S} = \mathbf{r}^{-\frac{1}{2}}(\mathbf{1}_n - (\mathbf{r} + \mathbf{z}^*)\mathbf{Y}_A)(\mathbf{1}_n - (\mathbf{r} - \mathbf{z})\mathbf{Y}_A)^{-1}\mathbf{r}^{\frac{1}{2}} \qquad (6.1.7)$$

Thus, the existence of the **S** matrix depends on the existence of the inverse $(\mathbf{1}_n - (\mathbf{r} - \mathbf{z})\mathbf{Y}_A)^{-1}$. This is very simple to assure. For example, as $\mathbf{1}_n$, **r**, and **z** are diagonal matrices, existence of one coupling in $\mathbf{Y}_A$ makes the condition fulfilled. The only exception is for one-ports, $1 = (\mathbf{r} - \mathbf{z})\mathbf{Y}_A$ and it is trivial that scattering matrix exists in this case.

6.2   Linearity, Passivity, and Causality

It is proven that a linear, passive n-port is always causal [Youla59].
     Meaning of causality is that the reason precedes the consequence in time.
     A possible definition of linearity is:

$$\mathbf{i}(t) = \int_{-\infty}^{\infty} \mathbf{H}(t - \tau)\mathbf{v}(\tau)d\tau \qquad (6.2.1)$$

The upper integral limit contains violation of causality.
Time-invariance has also been assumed. For causality,

$$\mathbf{H}(t - \tau) = 0 \text{ for } \tau \in [t, \infty), \quad t > t_0 \qquad (6.2.2)$$

Our goal is to prove (6.2.2) when the excitations are arbitrary. With the notations of Fig. 6.1.1,

$$\mathbf{v}_A = \mathbf{v} + \mathbf{ri} \tag{6.2.3}$$

where the entries of $\mathbf{r}$ are strictly positive. Passivity is defined as

$$\text{Re} \int_{t_0}^{t} \mathbf{i}^{*T}(\tau)\mathbf{v}(\tau)d\tau \geq 0 \text{ for all } t \in [t_0, \infty) \tag{6.2.4}$$

if the circuit was in zero-energy state at t_0 [Kuh67]. We assume voltage sources in the places of v_A and we assume that

$$\mathbf{v}_A(\tau) = 0_n \text{ for } \tau < t \tag{6.2.5}$$

Then

$$\mathbf{v}(\tau) = \mathbf{v}_A - \mathbf{ri} \tag{6.2.6}$$

Substituting (6.2.6) into (6.2.4):

$$\text{Re} \int_{t_0}^{t} \mathbf{i}^{*T}(\tau)(-\mathbf{ri}(\tau))d\tau \geq 0 \tag{6.2.7}$$

Consequently

$$\mathbf{i}(\tau) = 0_n \text{ for } \tau \in [t_0, t] \tag{6.2.8}$$

Take $t_0 = -\infty$ and from (6.2.1),

$$\mathbf{i}(t) = \int_{-\infty}^{t} \mathbf{H}(t-\tau)\mathbf{v}(\tau)d\tau + \int_{t}^{\infty} \mathbf{H}(t-\tau)\mathbf{v}(\tau)d\tau \tag{6.2.9}$$

The first term on the right side is zero because $\mathbf{v}(\tau)$ is zero for $\tau \in (-\infty, t]$. As in the second term, $\mathbf{v}(\tau) \neq 0_n$ for $\tau \in [t, \infty)$, and the same holds for arbitrary $v_A(\tau)$ *for* $\tau \in [t, \infty)$ thus arbitrary $\mathbf{v}(\tau)$,

$$\mathbf{H}(t-\tau) = 0_n \text{ for } \tau \in [t, \infty) \tag{6.2.10}$$

Nonlinear extensions are found in [Sandberg65].

6.3   Summary

Power matching has many consequences in circuit theory. Two of them have been selected here: Existence of the scattering matrix and causality. Note that here, not simply the known proofs have been repeated. Instead, based on the knowledge that can be learnt from the existing approaches, our own views have also been included.

References

[Youla59]   D.C. Youla, L.J. Castriota, H.J. Carlin, Bounded real scattering matrices and the foundations of linear passive network theory. IRE Trans. Circuit Theory **CT-6**, 102–124 (1959)

[Sandberg65]   I.W. Sandberg, Conditions for the causality of nonlinear operators defined on a function space. Quart. Appl. Math. **XXIII**(1), 87–91 (1965)

[Kuh67]   E.S. Kuh, R.A. Rohrer, Theory of Linear Active Networks (Holden-Day, Inc., SanFrancisco, 1967)

Chapter 7
Special Cases: Describing Functions and Weakly Nonlinear Case (Ladvánszky)

7.1 Power Matching in Nonlinear, Dynamic Circuits Characterized by Describing Functions

7.1.1 Historical Overview

See please Sect. 1.2.

7.1.2 Optimum Load Characteristics in Terms of Admittance Describing Functions

Please look at Fig. 3.1.1 again. Now we allow that N_G, in addition to nonlinear resistors, contains capacitors, inductors, memristors [Chua71], and periodic current and voltage sources. We assume that, because of appropriate time periods of the source signals and bandpass properties of the circuit, port currents and voltages are sinusoidal:

$$\underline{i} = \mathrm{Re}\!\left(\underline{I}e^{j\omega_0 t}\right) \quad \underline{v} = \mathrm{Re}\!\left(\underline{V}e^{j\omega_0 t}\right) \tag{7.1.2.1}$$

where $\underline{I}$ and $\underline{V}$ are the column matrices of the port currents and voltages, respectively. Thus, N_G can be characterized by the set of admissible signal pairs $S = \left\{\underline{I}, \underline{V}\right\}$.

Let a resistive 2n-port be characterized by the set of admissible signal pairs $S_R = \left\{\underline{I_R}, \underline{V_R}\right\}$ where

© The Author(s), under exclusive licence to Springer Nature Switzerland AG 2019
J. Ladvánszky, *Theory of Power Matching*, SpringerBriefs in Electrical and
Computer Engineering, https://doi.org/10.1007/978-3-030-16631-1_7

$$
\underline{I}_R = \begin{bmatrix} \operatorname{Re} I_1 \\ \operatorname{Im} I_1 \\ \operatorname{Re} I_2 \\ \operatorname{Im} I_2 \\ \vdots \\ \operatorname{Re} I_n \\ \operatorname{Im} I_n \end{bmatrix} \qquad \underline{V}_R = \begin{bmatrix} \operatorname{Re} V_1 \\ \operatorname{Im} V_1 \\ \operatorname{Re} V_2 \\ \operatorname{Im} V_2 \\ \vdots \\ \operatorname{Re} V_n \\ \operatorname{Im} V_n \end{bmatrix} \tag{7.1.2.2}
$$

and the reference directions are the same as those for N_G. Let the power of this port and that of N_G be identical:

$$
\operatorname{Re}\left(\underline{I}^{*T}\underline{V}\right) = \underline{I}_R{}^T \underline{V}_R \tag{7.1.2.3}
$$

and Theorem 3.2.2 can be applied for the resistive 2n-port and thus for the dynamic n-port as well. If the conditions of Theorem 3.2.2 hold, the dynamic n-port has a Norton equivalent:

$$
\underline{I}_S - \underline{f}(\underline{V}) = \underline{I} \tag{7.1.2.4}
$$

As we assume sinusoidal currents and voltages, describing functions for sinusoidal excitations can be determined ([Gelb68], p. 42). Similarly, as in [Mazumder77], we introduce the admittance describing functions $Y_{Gk} = G_{Gk} + jB_{Gk}$, $Y_{Lk} = G_{Lk} + jB_{Lk}$ corresponding to the kth port ($k=1,2, \ldots$ n) of the generator and the load, respectively.

$$
f_k(\underline{V}) = V_k Y_{Gk}(\underline{V}) \tag{7.1.2.5}
$$

$$
I_k = V_k Y_{Lk}(\underline{V}) \tag{7.1.2.6}
$$

In the following, real and imaginary parts will be denoted by indices 1 and 2, except at admittances:

$$
f_k(\underline{V}) = V_{k1}G_{Gk} - V_{k2}B_{Gk} + j(V_{k1}B_{Gk} + V_{k2}G_{Gk}) \tag{7.1.2.7}
$$

$$
I_k = V_{k1}G_{Lk} - V_{k2}B_{Lk} + j(V_{k1}B_{Lk} + V_{k2}G_{Lk}) \tag{7.1.2.8}
$$

Power dissipated by the load is:

$$
P(\underline{V}) = \sum_{k=1}^{n} \left[\left(I_{Sk1} - f_{k1}(\underline{V})\right)V_{k1} + \left(I_{Sk2} - f_{k2}(\underline{V})\right)V_{k2} \right] \tag{7.1.2.9}
$$

Necessary conditions for the power extremum are:

$$\frac{\partial P}{\partial V_{k1}} = 0 \quad \frac{\partial P}{\partial V_{k2}} = 0 \tag{7.1.2.10}$$

(7.1.2.4–7.1.2.10) yield for power extremum equations containing characteristics of the generator and the load:

$$V_{k1}G_{Lk} - V_{k2}B_{Lk} = V_{k1}G_{Gk} + V_{k2}B_{Gk} + \sum_{m=1}^{n} \left(V_{m1}^2 + V_{m2}^2\right) \frac{\partial G_{Gm}}{\partial V_{k1}} \tag{7.1.2.11}$$

$$V_{k1}B_{Lk} + V_{k2}G_{Lk} = -V_{k1}B_{Gk} + V_{k2}G_{Gk} + \sum_{m=1}^{n} \left(V_{m1}^2 + V_{m2}^2\right) \frac{\partial G_{Gm}}{\partial V_{k2}} \tag{7.1.2.12}$$

This system of equation is solved for G_{Lk} and B_{Lk}:

$$G_{Lk} = G_{Gk} + \frac{1}{V_{k1}^2 + V_{k2}^2} \sum_{m=1}^{n} \left(V_{m1}^2 + V_{m2}^2\right) \left(V_{k1}\frac{\partial G_{Gm}}{\partial V_{k1}} + V_{k2}\frac{\partial G_{Gm}}{\partial V_{k2}}\right) \tag{7.1.2.13}$$

$$B_{Lk} = -B_{Gk} + \frac{1}{V_{k1}^2 + V_{k2}^2} \sum_{m=1}^{n} \left(V_{m1}^2 + V_{m2}^2\right) \left(-V_{k2}\frac{\partial G_{Gm}}{\partial V_{k1}} + V_{k1}\frac{\partial G_{Gm}}{\partial V_{k2}}\right)$$
$$\tag{7.1.2.14}$$

For the transfer to polar coordinates, we use the following identities:

$$\frac{\partial G_{Gm}}{\partial V_{k1}} = \frac{\partial G_{Gm}}{\partial |V_k|} \frac{V_{k1}}{|V_k|} - \frac{\partial G_{Gm}}{\partial \varphi_k} \frac{V_{k2}}{|V_k|^2} \tag{7.1.2.15}$$

$$\frac{\partial G_{Gm}}{\partial V_{k2}} = \frac{\partial G_{Gm}}{\partial |V_k|} \frac{V_{k2}}{|V_k|} + \frac{\partial G_{Gm}}{\partial \varphi_k} \frac{V_{k1}}{|V_k|^2} \tag{7.1.2.16}$$

where we introduced the following notations:

$$V_k = |V_k| e^{j\varphi_k} \tag{7.1.2.17}$$

For real and imaginary parts of the admittance describing functions of the optimum load, (7.1.2.13–7.1.2.16) yield

$$G_{Lk} = G_{Gk} + \frac{1}{|V_k|} \sum_{m=1}^{n} |V_m|^2 \frac{\partial G_{Gm}}{\partial |V_k|} \tag{7.1.2.18}$$

$$B_{Lk} = -B_{Gk} + \frac{1}{|V_k|^2} \sum_{m=1}^{n} |V_m|^2 \frac{\partial G_{Gm}}{\partial \varphi_k} \qquad (7.1.2.19)$$

Admittance describing functions can be defined in other way [Baranyi84], but as we did here is the simplest for optimum load.

For one ports, (7.1.2.18 and 7.1.2.19) are simplified as follows:

$$G_L = G_G + |V| \frac{\partial G_G}{\partial |V|} \qquad (7.1.2.20)$$

$$B_L = -B_G \qquad (7.1.2.21)$$

because for time-invariant circuits, the describing function is independent of φ.

Can we simplify (7.1.2.18 and 7.1.2.19) using differentiation with respect to a complex variable? The answer is negative:

Theorem 7.1.2.1 Describing functions are differentiable in Cauchy-Riemann sense if and only if the circuit is linear.

Proof We prove it for $f_k(\underline{V})$ in (7.1.2.5). Necessity is obvious. For proving sufficiency, we start from the relation between first harmonic currents and voltages:

$$\underline{f}(\underline{V}) = \frac{1}{2\pi} \int_{-\pi}^{\pi} \underline{F}\left(Re\ \underline{V}e^{j\omega_0 t}\right) e^{-j\omega_0 t} d\omega_0 t \qquad (7.1.2.22)$$

where $\underline{i} = \underline{F}(v)$ is the constitutive relation of the investigated n-port, it is nonlinear. Essential that we neglected the DC and the higher harmonic current components:

$$\int_{-\pi}^{\pi} \underline{F}\left(Re\ \underline{V}e^{j\omega_0 t}\right) e^{-jl\omega_0 t} d\omega_0 t = 0, \quad l = 0,2,3,4, \ldots \qquad (7.1.2.23)$$

The Cauchy-Riemann relations for the kth current component are as follows:

$$\frac{\partial f_{k1}}{\partial V_{m1}} - \frac{\partial f_{k2}}{\partial V_{m2}} = 0 \qquad (7.1.2.24)$$

$$\frac{\partial f_{k1}}{\partial V_{m2}} + \frac{\partial f_{k2}}{\partial V_{m1}} = 0 \qquad (7.1.2.25)$$

Application of the Cauchy-Riemann relations for (7.1.2.22) yields

$$\int_{-\pi}^{\pi} Df_k\left(\mathrm{Re}\ \underline{V}e^{j\omega_0 t}\right) \cos\left(2\omega_0 t\right) d\omega_0 t = 0 \qquad (7.1.2.26)$$

$$\int_{-\pi}^{\pi} Df_k\left(\mathrm{Re}\ \underline{V}e^{j\omega_0 t}\right) \sin\left(2\omega_0 t\right) d\omega_0 t = 0 \qquad (7.1.2.27)$$

where D denotes the Fréchet differential operator ([Ortega70], p. 61).

From the derivatives of (7.1.2.23), and (7.1.2.26 and 7.1.2.27) we get

$$\int_{-\pi}^{\pi} Df_k\left(\mathrm{Re}\ \underline{V}e^{j\omega_0 t}\right) \cos\left(l\omega_0 t\right) d\omega_0 t = 0 \quad l = 1,2,3, \ldots \qquad (7.1.2.28)$$

$$\int_{-\pi}^{\pi} Df_k\left(\mathrm{Re}\ \underline{V}e^{j\omega_0 t}\right) \sin\left(l\omega_0 t\right) d\omega_0 t = 0 \quad l = 1,2,3, \ldots \qquad (7.1.2.29)$$

(7.1.2.28 and 7.1.2.29) hold if and only if

$$Df_k\left(\mathrm{Re}\underline{V}e^{j\omega_0 t}\right) = \mathrm{const.} \qquad (7.1.2.30)$$

or in other words, $f_k(.)$ is a linear map.

Application of sinusoidal input describing functions is an approximate characterization of the circuit, because (7.1.2.23) is never fulfilled exactly. Thus, the theorem above has a theoretical value; it points out that neglection of harmonic currents does not necessarily result in linearity of the investigated circuit, but it does together with assuming differentiability.

7.1.3   Special Case: Linear n-ports

Subject of this chapter is to show how the optimum load for linear circuits will follow from nonlinear one, the results known from literature for linear circuits. After that we will also show how the conditions of power matching for linear circuits will follow from the nonlinear ones.

We start from the Norton equivalent of the linear generator comprising current sources and a multiport characterized by the admittance matrix $\underline{\underline{Y}}_G$ according to Fig. 3.2.2. In this case, Eq. (7.1.2.5) will look like

$$f_k(\underline{V}) = \sum_{l=1}^{n} Y_{Gkl} V_l \qquad (7.1.3.1)$$

where Y_{Gkl} is an entry of the admittance matrix $\underline{\underline{Y}}_G$. Interconnection between the admittance describing function Y_{Gk} defined by (7.1.2.5) and the entries of the admittance matrix is the following:

$$Y_{Gk} = Y_{Gkk} + \sum_{l \neq k}^{n} Y_{Gkl} \frac{V_l}{V_k} \qquad (7.1.3.2)$$

The admittance describing function and the entries of the admittance matrix are distinguished by the different number of indices. The admittance describing function in (7.1.3.2) is separated into conductance and susceptance:

$$G_{Gk} = G_{Gkk} + \sum_{l \neq k}^{n} (G_{Gkl} \cos \varphi_{lk} - B_{Gkl} \sin \varphi_{lk}) \frac{|V_l|}{|V_k|} \qquad (7.1.3.3)$$

$$B_{Gk} = B_{Gkk} + \sum_{l \neq k}^{n} (G_{Gkl} \sin \varphi_{lk} + B_{Gkl} \cos \varphi_{lk}) \frac{|V_l|}{|V_k|} \qquad (7.1.3.4)$$

$$\varphi_{lk} = \varphi_l - \varphi_k \qquad (7.1.3.5)$$

To determine the describing functions of the load, the derivatives in Eqs. (7.1.2.18 and 7.1.2.19) should be calculated:

$$\left. \frac{\partial G_{Gm}}{\partial |V_k|} \right|_{m \neq k} = (G_{Gmk} \cos \varphi_{km} - B_{Gmk} \sin \varphi_{km}) \frac{1}{|V_m|} \qquad (7.1.3.6)$$

$$\left. \frac{\partial G_{Gm}}{\partial |V_k|} \right|_{m=k} = \sum_{l \neq m} (G_{Gml} \cos \varphi_{lm} - B_{Gml} \sin \varphi_{lm}) \left(-\frac{|V_l|}{|V_m|^2} \right) \qquad (7.1.3.7)$$

$$\left. \frac{\partial G_{Gm}}{\partial \varphi_k} \right|_{m \neq k} = -(G_{Gmk} \sin \varphi_{km} + B_{Gmk} \cos \varphi_{km}) \frac{|V_k|}{|V_m|} \qquad (7.1.3.8)$$

$$\left. \frac{\partial G_{Gm}}{\partial \varphi_k} \right|_{m=k} = \sum_{l \neq m} (G_{Gml} \sin \varphi_{lm} + B_{Gml} \cos \varphi_{lm}) \frac{|V_l|}{|V_m|} \qquad (7.1.3.9)$$

where Eq. (7.1.3.5) was applied. Now we substitute Eqs. (7.1.3.3–7.1.3.9) into (7.1.2.18 and 7.1.2.19). For easier understanding, we show on the left margin, the corresponding parts of (7.1.2.18 and 7.1.2.19).

$$(G_{Gk}) \quad G_{Lk} = G_{Gkk} + \sum_{l \neq k}^{n} (G_{Gkl} \cos \varphi_{lk} - B_{Gkl} \sin \varphi_{lk}) \frac{|V_l|}{|V_k|} - \left(\frac{1}{|V_k|} |V_k|^2 \frac{\partial G_{Gk}}{\partial |V_k|} \right)$$

$$- \sum_{l \neq k}^{n} (G_{Gkl} \cos \varphi_{lk} - B_{Gkl} \sin \varphi_{lk}) \frac{|V_l|}{|V_k|} + \left(\frac{1}{|V_k|} \sum_{m \neq k}^{n} |V_m|^2 \frac{\partial G_{Gm}}{\partial |V_k|} \right)$$

$$+ \frac{1}{|V_k|} \sum_{m \neq k}^{n} (G_{Gmk} \cos \varphi_{km} - B_{Gmk} \sin \varphi_{km}) |V_m| \qquad (7.1.3.10)$$

It is obvious that the second and third terms on the right will vanish. Applying the relation $\varphi_{km} = -\varphi_{mk}$, the fourth term will be reformulated:

$$G_{Lk} = G_{Gkk} + \sum_{m \neq k}^{n} (G_{Gmk} \cos \varphi_{mk} + B_{Gmk} \sin \varphi_{mk}) \frac{|V_m|}{|V_k|} \qquad (7.1.3.11)$$

The same for the susceptance:

$$(-B_{Gk}) \quad B_{Lk} = -B_{Gkk} - \sum_{l \neq k}^{n} (G_{Gkl} \sin \varphi_{lk} + B_{Gkl} \cos \varphi_{lk}) \frac{|V_l|}{|V_k|}$$

$$+ \left(\frac{\partial G_{Gk}}{\partial \varphi_k} \right) + \sum_{l \neq k}^{n} (G_{Gkl} \sin \varphi_{lk} + B_{Gkl} \cos \varphi_{lk}) \frac{|V_l|}{|V_k|} - \left(\frac{1}{|V_k|} \sum_{m \neq k}^{n} |V_m|^2 \frac{\partial G_{Gm}}{\partial \varphi_k} \right)$$

$$- \sum_{m \neq k}^{n} (G_{Gmk} \sin \varphi_{km} + B_{Gmk} \cos \varphi_{km}) \frac{|V_m|}{|V_k|} \qquad (7.1.3.12)$$

Simplifying similarly to the conductance:

$$B_{Lk} = -B_{Gkk} + \sum_{m \neq k}^{n} (G_{Gmk} \sin \varphi_{mk} - B_{Gmk} \cos \varphi_{mk}) \frac{|V_m|}{|V_k|} \qquad (7.1.3.13)$$

Relation of the admittance describing function of the load to its admittance matrix is

$$Y_{Lk} = Y_{Lkk} + \sum_{m \neq k}^{n} Y_{Lkm} \frac{V_m}{V_k} \qquad (7.1.3.14)$$

On the other hand, we substitute (7.1.3.11) and (7.1.3.13) into $Y_{Lk} = G_{Lk} + jB_{Lk}$:

$$Y_{Lk} = Y_{Gkk}^* + \sum_{m \neq k}^{n} Y_{Gmk}^* \frac{V_m}{V_k} \tag{7.1.3.15}$$

Comparison between the last two equations results in

$$\sum_{m}^{n} Y_{Lkm} \frac{V_m}{V_k} = \sum_{m}^{n} Y_{Gmk}^* \frac{V_m}{V_k} \tag{7.1.3.16}$$

As the development is valid for arbitrary voltages, with trivial exceptions, relation between the admittance matrices of the load and the generator is $\underline{\underline{Y}}_L = \underline{\underline{Y}}_G^{*T}$ that is identical to the relation known from the literature.

The condition of power maximum for linear n-ports is derived from the condition A.3 reformulated for describing functions. The function $\underline{h}_R(\underline{V}_R)$ should be strictly monotonically increasing, where

$$\underline{h}_R = \underline{f}_R + \underline{\underline{J}}_R^T \underline{V}_R \tag{7.1.3.17}$$

and $\underline{I}_R = \underline{f}_R(\underline{V}_R)$ is defined by (7.1.2.2).

Considering (7.1.2.2), (7.1.3.1), and (7.1.3.2), odd and even components of the port current are:

$$f_{R,2k-1} = \sum_{l=1}^{n} (G_{Gkl} V_{l1} - B_{Gkl} V_{l2}) \tag{7.1.3.18}$$

$$f_{R,2k} = \sum_{l=1}^{n} (G_{Gkl} V_{l2} + B_{Gkl} V_{l1}) \tag{7.1.3.19}$$

Thus, the derivatives in (7.1.3.17) are

$$J_{R,2m-1,2l-1} = \frac{\partial f_{R,2m-1}}{\partial V_{R,2l-1}} = G_{Gml} \tag{7.1.3.20}$$

$$J_{R,2m-1,2l} = \frac{\partial f_{R,2m-1}}{\partial V_{R,2l}} = -B_{Gml} \tag{7.1.3.21}$$

$$J_{R,2m,2l-1} = \frac{\partial f_{R,2m}}{\partial V_{R,2l-1}} = B_{Gml} \tag{7.1.3.22}$$

$$J_{R,2m,2l} = \frac{\partial f_{R,2m}}{\partial V_{R,2l}} = G_{Gml} \tag{7.1.3.23}$$

Thus, the entry of (7.1.3.17) of odd and even index can be written as

$$h_{R,2k-1} = \sum_{l=1}^{n}(G_{Gkl}V_{l1} - B_{Gkl}V_{l2}) + \sum_{l=1}^{n}(G_{Glk}V_{l1} + B_{Glk}V_{l2}) \qquad (7.1.3.24)$$

$$h_{R,2k} = \sum_{l=1}^{n}(G_{Gkl}V_{l2} + B_{Gkl}V_{l1}) + \sum_{l=1}^{n}(G_{Glk}V_{l2} - B_{Glk}V_{l1}) \qquad (7.1.3.25)$$

Iff $\underline{h}_R(\underline{V}_R)$ is strictly monotonically increasing, then $\underline{V}_R^T \underline{h}_R > 0$ for arbitrary $\underline{V}_R$. Detailing this yields the following inequality:

$$\sum_{k=1}^{n}V_{k1}\sum_{l=1}^{n}V_{l1}(G_{Gkl} + G_{Glk}) + V_{l2}(B_{Glk} - B_{Gkl})+$$

$$+ \sum_{k=1}^{n}V_{k2}\sum_{l=1}^{n}V_{l1}(B_{Gkl} - B_{Glk}) + V_{l2}(G_{Gkl} + G_{Glk}) > 0 \qquad (7.1.3.26)$$

This inequality can be written simplified as $\underline{V}^{*T}\left(\underline{\underline{Y}}_G + \underline{\underline{Y}}_G^{*T}\right)\underline{V} > 0$ for all $\underline{V}$, that is, $\underline{\underline{Y}}_G + \underline{\underline{Y}}_G^{*T}$ is positive definite, according to (Desoer73).

7.1.4 Optimum Load of Generators in Terms of Scattering Describing Functions

Assume that the generator can be characterized by wave parameters:

$$\underline{a}_S + \underline{f}(\underline{b}_G) = \underline{a}_G \qquad (7.1.4.1)$$

where $\underline{a}_G$ and $\underline{b}_G$ denote the column matrices of the complex amplitudes of wave parameters travelling from the generator towards the load and back, respectively, and $\underline{a}_S$ is the independent wave parameter matrix. We assume nonlinear reflection coefficients depending on $\underline{b}_G$ (Fig. 7.1.4.1).

Determine the relation between $\underline{a}_L$ and $\underline{b}_L$ so that the dissipated power on the load P is maximum:

$$P = \underline{a}^{*T}\underline{a} - \underline{b}^{*T}\underline{b} \qquad (7.1.4.2)$$

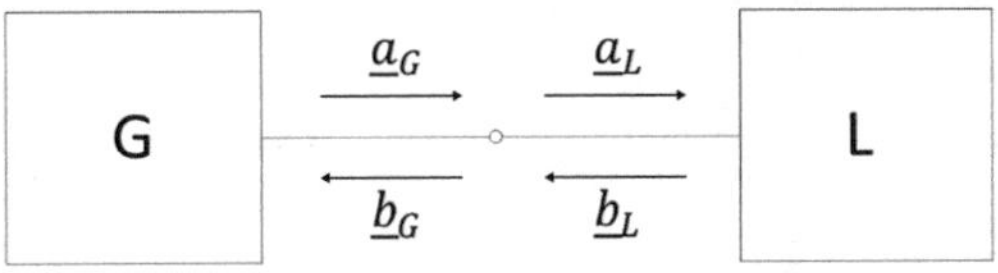

Fig. 7.1.4.1 Wave parameter description of a nonlinear generator and load

Because of the relations $\underline{a}_G = \underline{a}_L$ and $\underline{b}_G = \underline{b}_L$, temporarily we leave the index out. In the following, we repeat calculations of Sect. 7.1.2. We calculate for one-port only. The generator and the load are characterized by the scattering describing function $\Gamma_G(b)$ and the inverse scattering describing function $Q_L(b)$, respectively.

$$f(b) = b\Gamma_G(b) \tag{7.1.4.3}$$

$$a = bQ_L(b) \tag{7.1.4.4}$$

Real and imaginary parts will be denoted by 1 and 2 written in the index, respectively. Conditions of maximum power $\frac{\partial P}{\partial b_1} = 0$ and $\frac{\partial P}{\partial b_2} = 0$ yield

$$-(b_1 Q_{L1} - b_2 Q_{L2})\Gamma_{G1} - (b_1 Q_{L2} + b_2 Q_{L1})\Gamma_{G2}$$
$$-\left(b_1^2 + b_2^2\right)\left(Q_{L1}\frac{\partial \Gamma_{G1}}{\partial b_1} + Q_{L2}\frac{\partial \Gamma_{G2}}{\partial b_1}\right) - b_1 = 0 \tag{7.1.4.5}$$

$$+(b_1 Q_{L1} - b_2 Q_{L2})\Gamma_{G2} - (b_1 Q_{L2} + b_2 Q_{L1})\Gamma_{G1}$$
$$-\left(b_1^2 + b_2^2\right)\left(Q_{L1}\frac{\partial \Gamma_{G1}}{\partial b_2} + Q_{L2}\frac{\partial \Gamma_{G2}}{\partial b_2}\right) - b_2 = 0 \tag{7.1.4.6}$$

After rearranging, we apply identities similar to (7.1.2.15 and 7.1.2.16):

$$\frac{\partial \Gamma}{\partial b_1} = \frac{\partial \Gamma}{\partial |b|}\frac{b_1}{|b|} - \frac{\partial \Gamma}{\partial \varphi}\frac{b_2}{|b|^2} \tag{7.1.4.7}$$

$$\frac{\partial \Gamma}{\partial b_2} = \frac{\partial \Gamma}{\partial |b|}\frac{b_2}{|b|} - \frac{\partial \Gamma}{\partial \varphi}\frac{b_1}{|b|^2} \tag{7.1.4.8}$$

The result is

$$\frac{1}{|Q_L|} = |\Gamma_G| + |b|\frac{\partial |\Gamma_G|}{\partial |b|} \tag{7.1.4.9}$$

$$\varphi_Q = \varphi_G \tag{7.1.4.10}$$

$$\Gamma_G = |\Gamma_G|e^{j\varphi_G}, Q_L = |Q_L|e^{j\varphi_Q} \tag{7.1.4.11}$$

The result above specifies a load with inverse scattering describing function $Q_L(b)$ that dissipates maximum power at arbitrary $\underline{a}_S$. If $\underline{a}_S$ is fixed, then only one point of the optimum load is given that can be described by a level independent reflection coefficient Γ_L:

$$|\Gamma_L| = |\Gamma_G| + |b|\frac{\partial |\Gamma_G|}{\partial |b|} \tag{7.1.4.12}$$

Fig. 7.1.4.2 Comparison between the admittance- and the scattering describing function solutions of the power matching problem

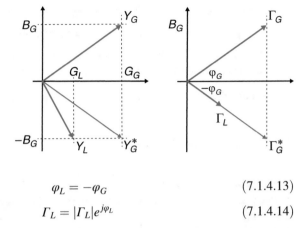

$$\varphi_L = -\varphi_G \qquad\qquad (7.1.4.13)$$

$$\Gamma_L = |\Gamma_L| e^{j\varphi_L} \qquad\qquad (7.1.4.14)$$

We can see that the relation $\Gamma_L = \Gamma_G^*$ known for linear circuits is not valid for nonlinear circuits, but it is a special case.

Let us compare the solutions of the power matching problem using admittance- (Sect. 7.1.2) and scattering describing functions. The similarity is obvious between (7.1.2.20 and 7.1.2.21) and (7.1.4.12 and 7.1.4.13), shown in Fig. 7.1.4.2, but validity is completely different.

Equations (7.1.2.20 and 7.1.2.21) are valid at sinusoidal excitations of arbitrary amplitude, while (7.1.4.12 and 7.1.4.13) are valid for fixed amplitude. For the existence of power maximum, other conditions are also needed. For admittance describing functions, as we mentioned, Theorems 3.2.1 and 3.2.2 are needed, extended according to Eq. (7.1.2.2). As we have seen, basis of these theorems is that power can be written in the form of an inner product, see please Eqs. (3.2.10) and (7.1.2.3). In case of the wave parameter characterization, power expression given by (7.1.4.2) does not have the properties of an inner product [Korn-Korn, 1975, Section 14.2.6]. For this reason, Theorems 3.2.1 and 3.2.2 are not valid for wave parameter characterization. Instead, the following statement is valid:

Theorem 7.1.4.1 If the nonlinear generator described by (7.1.4.1) has a power (7.1.4.2) maximum for arbitrary $\underline{a}_S$, then the inverse scattering describing function of the load Q_L is given by (7.1.4.9 and 7.1.4.10).

Note that property P5 of the optimum load also holds for wave parameter characterization:

P.5′ Wave parameters of the load $\underline{a}_L$ and $\underline{b}_L$, disconnecting the load from the source, does not depend on any P_k parameters of the generator:

$$\frac{\partial \underline{a}_L}{\partial P_k} = 0 \qquad \frac{\partial \underline{b}_L}{\partial P_k} = 0 \qquad\qquad (7.1.4.15)$$

Repeating ideas in Sect. 3.2 yields the following Theorem:

Theorem 7.1.4.2 If the sinusoidal generator characterized by the function $\underline{a}_G\left(\underline{b}_{G}, \underline{P}\right)$ that is differentiable has a load at arbitrary values of $\underline{P}$ dissipating maximum power, characterizable by the property P.5', then the generator characteristic can be described by Eq. (7.1.4.1) where $\underline{a}_S = \underline{a}_S\left(\underline{P}\right)$.

Proof Part of the proof of Theorem 3.2.2, using wave parameters.

7.1.5   Summary, Remarks

Power matching applying describing functions has been investigated. Closed form characteristics of the optimum load have been given for admittance and scattering describing functions. Significance of these results is that the formulae can be directly used in circuit design.

7.2   Maximum Power Transfer in Weakly Nonlinear Circuits (Ladvánszky)

7.2.1   Introduction

Problem of maximum power transfer has been investigated in the case when the conductance of the generator is given by its third-order Volterra series. We prove that in this case, the load can be characterized also by a third-order Volterra series. Closed-form relations are given for the Volterra series of the optimum load and its frequency domain representation. Our results are based on the mathematical identity (7.2.2.7).

Given the source circuit N_G shown in Fig. 7.2.1.1 comprising nonlinear circuit elements and independent generators with variable parameters, the problem is to find the characteristics of the load circuit N_L absorbing maximum power for arbitrary values of the variable parameters.

Several cases of the above problem differing in the assumptions on the source Ns have been studied in the literature: resistive sources [Wyatt83B, Ladvánszky86A], general nonlinear dynamic sources [Wyatt83A], and nonlinear dynamic sources characterized by sinusoidal input describing functions [Ladvánszky87]. The reverse problem when the source characteristic is determined from the given load characteristic has also been solved [Ladvánszky86B].

Fig. 7.2.1.1 The maximum power transfer problem: Given the generator circuit N_G, it is to find N_L absorbing maximum power

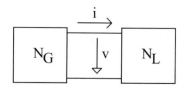

Fig. 7.2.1.2 The Norton
equivalent of the generator,
in parallel with the load

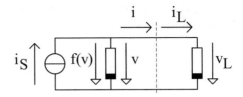

In the case of mixers and power amplifiers driven by a sum of sinusoids, maximum output power is often desirable. However, this problem cannot be treated using the abovementioned models of N_G. In several practical applications the range of excitations is limited; thus, Volterra series can be used advantageously.

In this section we assume a one-port source characterized by truncated Volterra series (Fig. 7.2.1.2). Excitations of arbitrary waveform are allowed. The maximum power transfer problem is solved by generalizing the method presented by Rohrer for the case of linear circuits [Rohrer65, Rohrer68]. Closed form expressions are derived for the Volterra kernels of the load absorbing maximum power in Sect. 7.2.4.

7.2.2   Notations and Assumptions

We assume a Norton-like source shown in Fig. 7.2.1.2 characterized by the following relation:

$$i(t) = i_S(t) - f[v(t)] \qquad (7.2.2.1)$$

where $i(t)$ denotes the port current, $v(t)$ denotes port voltage, and $i_S(t)$ denotes the generator waveform, and t is time. Generator nonlinearity is denoted by the time invariant operator $f(.)$. Current and voltage waveforms are square integrable in the interval $-\infty < t < \infty$. We determine the maximum of the net energy ε. Net energy is defined as follows [Rohrer68]:

$$\varepsilon = \int_{-\infty}^{\infty} v(t)i(t)dt \qquad (7.2.2.2)$$

And we introduce the notation:

$$\int_{-\infty}^{\infty} v(t)i(t)dt = \langle v, i \rangle \qquad (7.2.2.3)$$

The integral is finite due to the Schwartz inequality [Rohrer68]. Existence conditions for the power maximum and the Norton equivalent are found in the literature [Wyatt83A, Ladvánszky86A].

The generator is assumed weakly nonlinear in sense that $f(.)$ can be expressed by truncated Volterra series [Chua79]:

$$f[v_1(t), \ldots v_N(t)] = \sum_{n=1}^{N} f_n[v_1(t), \ldots v_n(t)] \tag{7.2.2.4}$$

$$f_n[v_1(t), \ldots v_n(t)] = \int_{-\infty}^{\infty} \cdots \int_{-\infty}^{\infty} h_n(\tau_1, \ldots \tau_n) \prod_{k=1}^{n} [v_k(t - \tau_k) d\tau_k] \tag{7.2.2.5}$$

The kernel $h_n(\tau_1, \tau_2, \ldots \tau_n)$ is denoted by the h_n shorthand notation:

$$f_n = h_n * \left(\overset{1}{v_1}, \overset{n}{\cdots . v_n} \right) \tag{7.2.2.6}$$

where * denotes convolution. In our investigations, the following identity has an important role:

$$\langle \alpha, h_n * (v_1, \ldots v_k, \ldots v_n) \rangle = \left\langle v_k, {}_kh^k n * \left(\overset{1}{v_1}, \ldots \overset{k}{\alpha}, \ldots v_n^n \right) \right\rangle \tag{7.2.2.7}$$

where $\alpha(t)$ is square integrable, and the left subscript denotes—sign in the argument:

$${}_kh_n = h_n(\tau_1 \ldots - \tau_k \ldots \tau_n) \tag{7.2.2.8}$$

The identity (7.2.2.7) has been proven for first-order kernel [Rohrer68]. For higher orders, the proof is similar:

$$\langle \alpha, h_n * (v_1, \ldots v_k, \ldots v_n) \rangle = \int_{-\infty}^{\infty} \alpha(t) \int_{-\infty}^{\infty} \cdots \int_{-\infty}^{\infty} h_n(\tau_1, \ldots \tau_n) \prod_{m=1}^{n} [v_m(t - \tau_m) d\tau_m] dt \Bigg|_{t - \tau_k = x} =$$

$$= - \int_{-\infty}^{\infty} v_k(x) \int_{-\infty}^{\infty} \cdots \int_{-\infty}^{\infty} h_n(\tau_1, \ldots, \tau_k, \ldots \tau_n) \prod_{\substack{m=1 \\ m \neq k}}^{n} [v_m(x + \tau_k - \tau_m) d\tau_m] \alpha(x + \tau_k) d\tau_k dx \Bigg|_{\tau_k \to -\tau_k} =$$

$$= \int_{-\infty}^{\infty} v_k(x) \int_{-\infty}^{\infty} \cdots \int_{-\infty}^{\infty} h_n(\tau_1, \ldots, -\tau_k, \ldots \tau_n) \prod_{\substack{m=1 \\ m \neq k}}^{n} [v_m(x - \tau_k - \tau_m) d\tau_m] \alpha(x - \tau_k) d\tau_k dx \Bigg|_{\tau_k + \tau_m \to \tau_m} =$$

$$= \left\langle v_k, {}_kh^k n * \left(\overset{1}{v_1}, \ldots \overset{k}{\alpha}, \ldots v_n^n \right) \right\rangle$$

Theorem 7.2.2.1 In case of an inner product, the term not containing convolution can be exchanged by an arbitrary term in the convolution that is not the kernel, in such a way that the value of the inner product is not changed. To do that, the sign of the proper argument of the kernel should be changed according to (7.2.2.7 and 7.2.2.8).

7.2.3 Maximum Power Transfer in Linear Circuits, Time, and Frequency Domains

Recent papers dealing with maximum power transfer in linear, sinusoidally excited circuits have been listed in [Lin85]. Unfortunately, the papers listed there do not contain [Rohrer65, Rohrer68] dealing with non-sinusoidal excitations as well. Essence of references [Rohrer65, Rohrer68] has been summarized in this section.

Please consider the one-port generator in Fig. 7.2.1.2, that is built up with the current source of the waveform $i_S(t)$ and the passive one-port described by the following equation:

$$f(v) = h_1{}^*v \tag{7.2.3.1}$$

The task is to find the kernel h_{L1} of the load absorbing maximum net energy

$$\varepsilon = \langle v_L, i_L \rangle \tag{7.2.3.2}$$

For arbitrary waveform $i_S(t)$ of the current source, where $v_L(t)$ and $i_L(t)$ are the voltage and current waveforms of the load, respectively:

$$i_L = h_{L1}{}^*v_L \tag{7.2.3.3}$$

and from the Figure, $i_L = i$, $v_L = v$. The maximum energy and the corresponding waveforms are denoted by zero index:

$$\varepsilon_0 = \langle v_{L0}, i_{L0} \rangle \tag{7.2.3.4}$$

The optimum load has been found by varying the voltage:

$$v = v_0 + \alpha \tag{7.2.3.5}$$

The condition of the global maximum is:

$$\varepsilon_0 - \varepsilon \geq 0 \tag{7.2.3.6}$$

$$\langle v_0, i_0 \rangle - \langle v_0 + \alpha, i \rangle \geq 0 \tag{7.2.3.7}$$

for arbitrary $\alpha(t)$. Rearranging (7.2.3.7):

$$\langle v_0, i_0 - i \rangle - \langle \alpha, i \rangle \geq 0 \tag{7.2.3.8}$$

Applying the form of generator characteristic according to (7.2.2.1), (7.2.3.5) yields

$$i_0 - i = h_1{}^*(v_0 + \alpha) - h_1{}^* v_0 = h_1{}^* \alpha \tag{7.2.3.9}$$

Substituting (7.2.3.9) into (7.2.3.8) yields

$$\langle v_0, h_1{}^* \alpha \rangle - \langle \alpha, i_S - h_1{}^*(v_0 + \alpha) \rangle \geq 0 \tag{7.2.3.10}$$

The inequality (7.2.3.10) holds for arbitrary $\alpha(t)$ if and only if

$$\left\langle \alpha, {}_1 h_1{}^1 * v_0 - i_S + h_1 * v_0 \right\rangle = 0 \tag{7.2.3.11}$$

and

$$\langle \alpha, h_1{}^* \alpha \rangle \geq 0 \tag{7.2.3.12}$$

Theorem 7.2.3.1 Energy delivered by the linear, time invariant generator has maximum, if the one-port kernel in the Norton equivalent of the generator obeys (7.2.3.12).

Circuit equation in the operating point v_0 corresponding to the optimum is:

$$h_{L1}{}^* v_0 - i_S + h_1{}^* v_0 = 0 \tag{7.2.3.13}$$

Therefore, (7.2.3.11) is fulfilled if

$$h_{L1} = {}_1 h_1 \tag{7.2.3.14}$$

Theorem 7.2.3.2 Kernel of the maximum energy load of a linear, time-invariant generator can be derived from the kernel of the passive one-port in the Norton equivalent of the generator by changing the sign of the argument, for arbitrary, square integrable waveform of the generator.

The inequality (7.2.3.12) is called as energy absorbing condition. Double-sided Laplace transform of the equation (7.2.3.14) gives the well-known complex frequency domain power matching condition:

$$Y_L(s) = Y(-s) \tag{7.2.3.15}$$

where Y_L, Y, and s denote load and source admittances and the complex frequency, respectively.

Theorem 7.2.3.3 Admittance of the maximum energy load of a linear, time-invariant generator can be derived from the admittance of the passive one-port in the Norton equivalent of the generator by changing the sign of the argument, for arbitrary, square integrable waveform of the generator.

The load admittance in (7.2.3.15) cannot be realized in general, but it can be approximated by realizable one-ports over a finite interval of the $j\omega$ axis. This topic is found in the literature as interpolation by positive real functions or as the Nevanlinna-Pick problem in circuit theory.

7.2.4   Power Matching in Weakly Nonlinear Circuits

Now the generator shown in Fig. 7.2.1.2. is investigated, with the following characteristics:

$$f(v) = h_1{}^*v + h_2{}^*(v, v) + h_3{}^*(v, v, v) \qquad (7.2.4.1)$$

The problem is to find kernels h_{L1}, h_{L2}, h_{L3} absorbing maximum energy $\varepsilon = \langle v_L, i_L \rangle$, where

$$i_L = h_{L1}{}^*v_L + h_{L2}{}^*(v_L, v_L) + h_{L3}{}^*(v_L, v_L, v_L) \qquad (7.2.4.2)$$

and $v_L = v$, $i_L = i$. The problem is solved by generalization of the steps obtained for linear circuits in the previous section.

Maximum energy and corresponding quantities will be denoted by the index 0. The condition of the global maximum has been obtained as follows:

$$\varepsilon_0 - \varepsilon \geq 0 \qquad (7.2.4.3)$$

$$\langle v_0, i_0 - i \rangle - \langle a, i \rangle \geq 0 \qquad (7.2.4.4)$$

In the left term of (7.2.4.4), $i_0 - i$ can be expressed as

$$
\begin{aligned}
i_0 - i = h_1{}^*a + \\
+ h_2{}^*[(v_0, a) + (a, v_0) + (a, a)] + \\
+ h_3{}^*[(v_0, v_0, a) + (v_0, a, v_0) + (a, v_0, v_0)] + \\
+ h_3{}^*[(v_0, a, a) + (a, v_0, a) + (a, a, v_0)] + \\
+ h_3{}^*(a, a, a)
\end{aligned}
\qquad (7.2.4.5)
$$

From (7.2.4.4) and (7.2.2.1) we get:

$$\langle v_0, i_0 - i \rangle - \langle \alpha, i_S - f(v) \rangle \geq 0 \tag{7.2.4.6}$$

where $i_0 - i$ is given by (7.2.4.5), $f(v)$ is given by (7.2.4.1). After substitution we apply (7.2.2.7) and (7.2.2.8):

$$\varepsilon_0 - \varepsilon = -\langle \alpha, i_S \rangle +$$
$$+\langle \alpha, (_1h_1 + h_1) * v_0 \rangle +$$
$$+\langle \alpha, (_1h_2 +_2h_2 + h_2) * (v_0, v_0) \rangle +$$
$$+\langle \alpha, (_1h_3 +_2h_3 +_3h_3 + h_3) * (v_0, v_0, v_0) \rangle$$

$$+\langle \alpha, h_1{}^* \alpha \rangle +$$
$$+\langle \alpha, (_1h_2 + h_2) * (v_0, \alpha) \rangle$$
$$+\langle \alpha, (_1h_3 + h_3) * (v_0, v_0, \alpha) + (_2h_3 + h_3) * (v_0, v_0, \alpha) + (_3h_3 + h_3) * (v_0, \alpha, v_0) \rangle +$$

$$+\langle \alpha, h_2{}^*(\alpha, \alpha) \rangle +$$
$$+\langle \alpha, (_1h_3 + h_3) * (v_0, \alpha, \alpha) + h_3{}^*(\alpha, v_0, \alpha) + h_3{}^*(\alpha, \alpha, v_0) \rangle +$$

$$+\langle \alpha, h_3{}^*(\alpha, \alpha, \alpha) \rangle \geq 0$$

$$(7.2.4.7)$$

where h_1, h_2, and h_3 are specified in (7.2.4.1), and the left index is explained in (7.2.2.7).

To make it easier, terms are grouped according to the power of α.

Eq. (7.2.4.7) holds for the domain $\{\alpha(t)\}$ of the square integrable functions $\alpha(t)$. This domain is decomposed to subdomains in which the waveforms are identical, but the amplitudes are not:

$$\{\alpha_0(t)\} = \{\alpha(t) | \alpha(t) = x\alpha_0(t)\} \tag{7.2.4.8}$$

where $\alpha_0(t)$ is the function generating the subdomain, and x is an arbitrary, finite real number. The inequality (7.2.4.7) holds for arbitrary $\alpha_0(t)$ and x; thus, (7.2.4.7) can be rewritten as follows:

$$\varepsilon_0 - \varepsilon = a_1 x + a_2 x^2 + a_3 x^3 + a_4 x^4 \geq 0 \tag{7.2.4.9}$$

where

$$a_1 = -\langle \alpha, i_S \rangle +$$
$$+\langle \alpha, (_1h_1 + h_1) * v_0 \rangle +$$
$$+\langle \alpha, (_1h_2 +_2h_2 + h_2) * (v_0, v_0) \rangle +$$
$$+\langle \alpha, (_1h_3 +_2h_3 +_3h_3 + h_3) * (v_0, v_0, v_0) \rangle$$

$$(7.2.4.10)$$

and by comparing (7.2.4.7) and (7.2.4.9), a_2, a_3, and a_4 can also be expressed. (7.2.4.9) holds for any x. This is right for $|x| \ll 1$ if

$$a_1 = 0 \tag{7.2.4.11}$$

or, based on (7.2.4.10),

$$\begin{aligned} &-\langle \alpha, i_S \rangle + \\ &+\langle \alpha, ({}_1h_1 + h_1) * v_0 \rangle + \\ &+\langle \alpha, ({}_1h_2 + {}_2h_2 + h_2) * (v_0, v_0) \rangle + \\ &+\langle \alpha, ({}_1h_3 + {}_2h_3 + {}_3h_3 + h_3) * (v_0, v_0, v_0) \rangle = 0 \end{aligned} \tag{7.2.4.12}$$

For this case, (7.2.4.9) is modified as:

$$\varepsilon_0 - \varepsilon = a_2 x^2 + a_3 x^3 + a_4 x^4 \geq 0 \tag{7.2.4.13}$$

Rearranging (7.2.4.13):

$$a_4 x^2 \left[\left(x + \frac{a_3}{2a_4} \right)^2 + \frac{4a_2 a_4 - a_3^2}{4a_4^2} \right] \geq 0 \tag{7.2.4.14}$$

which holds for any x if:

$$a_4 \geq 0 \tag{7.2.4.15}$$

and

$$4a_2 a_4 - a_3^2 \geq 0 \tag{7.2.4.16}$$

The inequality (7.2.4.15) can be rewritten as:

$$\langle \alpha, h_3{}^*(\alpha, \alpha, \alpha) \rangle \geq 0 \tag{7.2.4.17}$$

This is the generalization of the "energy absorbing condition" mentioned by [Rohrer65, Rohrer68].

Theorem 7.2.4.1 Energy delivered by the generator described by third-order truncated Volterra series has a maximum, if the kernels of the one-port in the Norton equivalent of the generator, and that of the load, fulfill the conditions (7.2.4.11, 7.2.4.15, and 7.2.4.16).

The equation characterizing the circuit in Fig. 7.2.1.2 is the following:

$$-i_S + (h_{L1} + h_1) * v_0 +$$
$$+(h_{L2} + h_2) * (v_0, v_0) + \qquad (7.2.4.18)$$
$$+(h_{L3} + h_3) * (v_0, v_0, v_0) = 0$$

By comparison of (7.2.4.18) and (7.2.4.12), we get the generalization of the "adjoint matching condition" mentioned by Rohrer:

$$h_{L1} = {}_1h_1 \qquad (7.2.4.19)$$

$$h_{L2} = {}_1h_2 + {}_2h_2 \qquad (7.2.4.20)$$

$$h_{L3} = {}_1h_3 + {}_2h_3 + {}_3h_3 \qquad (7.2.4.21)$$

Matching conditions in the complex frequency domain have been obtained by Laplace transform of (7.2.4.19–7.2.4.21):

$$H_{L1} = {}_1H_1 \qquad (7.2.4.22)$$

$$H_{L2} = {}_1H_2 + {}_2H_2 \qquad (7.2.4.23)$$

$$H_{L3} = {}_1H_3 + {}_2H_3 + {}_3H_3 \qquad (7.2.4.24)$$

where the following notation has been introduced:

$$H_n(s_1, \ldots s_n) = \int\limits_{-\infty}^{\infty} \cdots \int\limits_{-\infty}^{\infty} h_n(\tau_1, \ldots \tau_n) \prod_{k=1}^{n} (e^{-s_k \tau_k} d\tau_k) \qquad (7.2.4.25)$$

Theorem 7.2.4.2 Necessary condition of the energy maximum of the generator described by third-order truncated Volterra series is that the load kernels are given as (7.2.4.19–7.2.4.21), the transfer functions are given in (7.2.4.22–7.2.4.24).

Extending (7.2.4.1) for higher order terms yields:

$$f(v) = \sum_{k=1}^{n} h_k^* \overset{1, \ \ldots \ k}{(v, \ldots v)} \qquad (7.2.4.26)$$

and the steps above can be repeated. Extending (7.2.4.9), condition for the global energy maximum is:

$$\varepsilon_0 - \varepsilon = \sum_{k=1}^{n+1} a_k x^k \geq 0 \qquad (7.2.4.27)$$

Necessary conditions for the existence of the energy maximum, expressed by the coefficients, are as follows:

$$a_1 = 0 \tag{7.2.4.28}$$

$$a_{n+1} \geq 0 \tag{7.2.4.29}$$

providing that n is odd. Equation (7.2.4.28) leads to the extension of the "adjoint matching condition" for arbitrary, odd n:

$$h_{Lk} = \sum_{m=1}^{k} {}_mh_k \tag{7.2.4.30}$$

where $k = 1, 2, ...n$. Eq. (7.2.4.30) can be simplified by introducing the symmetrized kernel as follows:

$$\overline{h_{Lk}} = k\left({}_1\overline{h_k}\right) \tag{7.2.4.31}$$

$$\overline{h_k} = \frac{1}{k!} \sum h_k(\tau_1, \ldots \tau_k) \tag{7.2.4.32}$$

and the summation covers all permutations of $\tau_1, \ldots \tau_k$ (without repetition).

Laplace transform of (7.2.4.30) and (7.2.4.31) yields the matching conditions in the complex frequency domain:

$$H_{Lk} = \sum_{m=1}^{k} {}_mH_k \tag{7.2.4.33}$$

$$\overline{H_{Lk}} = k\left({}_1\overline{H_k}\right) \tag{7.2.4.34}$$

Theorem 7.2.4.3 Necessary condition of the energy maximum of a generator described by n term Volterra series is that the kernels of the load are as in (7.2.4.30), transfer functions of the load are as in (7.2.4.33).

The conditions given by (7.2.4.13), (7.2.4.15), and (7.2.4.16) are necessary and sufficient conditions. But conditions (7.2.4.28) and (7.2.4.29) are necessary but not sufficient conditions. Thus, (7.2.4.19–7.2.4.24), or (7.2.4.30) and (7.2.4.33) can be applied only if somehow we are convinced that the maximum exists.

If the generator in Fig. 7.2.1.2. is resistive and

$$f(v) = Kv^n \tag{7.2.4.35}$$

with $K > 0$, and odd n, then the optimum load exists [Wyatt83B]. Comparison of the optimum load characteristics $i_L = nKv_L^n$ and (7.2.4.33) shows that our results are extensions of the results for resistive circuits.

Causality of the generator implies that

$$h_k(\tau_1, \ldots \tau_k) = 0 \quad \tau_m < 0, m = 1, \ldots k \qquad (7.2.4.36)$$

Equations (7.2.4.30) and (7.2.4.33) imply:

$$h_{Lk} = 0, \ \tau_m > 0 \qquad (7.2.4.37)$$

but for negative arguments, load kernels can be arbitrary. Consequently, optimum loads may not be causal. Equations (7.2.4.19–7.2.4.24), (7.2.4.30 and 7.2.4.31), (7.2.4.33 and 7.2.4.34) in general characterize noncausal circuits. Problem of realization is to approximate them by realizable circuits at a finite interval of the $j\omega$ axis.

Writing in closed form the characteristics of the optimum load has a great practical significance. Numerically, for nonlinear generators, the load absorbing maximum power can be found using such programs that has both nonlinear analysis and optimization options, and these two can be simultaneously applied. Our experience is that this is not always effective because the nonlinear analysis and optimization (calculation of the necessary derivatives) are approximations, and this is a source of various numerical problems.

7.2.5 Conclusions and Remarks

Closed-form formulae have been derived for characteristics of the load absorbing maximum power from generator characterized by truncated Volterra series. Results have been given in both time and complex frequency domains. Conditions for existence of the energy maximum have also been studied: Necessary and sufficient conditions have been given for third-order case and necessary conditions for nth-order case (n is odd).

We note that periodic signals are not square integrable in the double infinite time interval. This problem can be resolved by considering a finite but long portion of the periodic excitation signal, considering the relation between average power and net energy over the finite time interval. This is the reason why we can use the term "maximum power transfer."

Basic assumptions	domain	Generator characteristics	Characteristics of the optimum load		Notations	References
Linear generator	t	$i = i_S - h * v$	$i_L = h_L * v_L$	$h_L = 1\ h_1$	h_1 : first order kernel	[Rohrer65][1]
	s	$I = I_S - Y(s)V$	$I_L = Y_L * (s)V$	$Y_L(s) = Y(-s)$	$Y(s)$: admittance	[Rohrer68][1]
Nonlinear resistive generator	t s ω	$i = i_S - f(v)$	$i_L = g(v_L)$	$g(v_L) = f'(v_L)v_L$	$f' = \frac{df}{dv}$	[Wyatt83B]
Weakly nonlinear generator	t	$i = i_S - \sum_{k-1}^{N} h_k * \begin{pmatrix} 1 \ldots k \\ v, \ldots v \end{pmatrix}$	$i_L = \sum_{k-1}^{N} h_{Lk} * \begin{pmatrix} 1 \ldots k \\ v_L, \ldots v_L \end{pmatrix}$	$h_{Lk} = \sum_{m=1}^{k} m\ h_k$	h_k : k-th order kernel	This chapter
	s	$I = I_S - \sum_{k-1}^{N} H_k \prod_{m-1}^{k} V_m$	$I_L = \sum_{k-1}^{N} H_{Lk} \prod_{m-1}^{k} V_{Lm}$	$H_{Lk} = \sum_{m=1}^{k} m\ H_k$	H_k : k-the order transfer function	This chapter
Tuned nonlinear	ω	$I = I_S - V[G(V) + jB(V)]$	$I_L = V_L\left[G_L(V_L) + jB_L(V_L)\right]$	$G_L(V_L) = G(V_L) +$ $+V_L \frac{dG(V)}{dV}\Big\vert_{V = V_L}$ $B_L(V_L) = -B(V_L)$	$G(V) + jB(V)$: admittance describing function	[Ladvánszky87]
generator[2]	ω	$a = a_S - b\vert\Gamma(b)\vert e^{j\varphi(b)}$	$b_L = a_L\vert\Gamma_L(b_L)\vert e^{j\varphi_L(b_L)}$	$\vert\Gamma_L(b_L)\vert = \vert\Gamma_L(b_L)\vert +$ $+b_L \frac{d\vert\Gamma(b)\vert}{db}\Big\vert_{b-b_L}$ $\varphi_L(b_L) = -\varphi(b_L)$	a, b : wave-parameters $\Gamma(b)$: reflection describing function	[Ladvánszky87]
General non-linear generator	t	$i = i_S - f(v)$	$i_L = g(V_L)$	$g(v_L) =$ $= \left[Df(v)\right]^{adj}\Big\vert_{v=v_L} v_L$	D : Gateaux differential-operator adj : adjoint operator	[Wyatt83A]

[1]These papers apply impedance formalism. [2]One phase can be freely chosen in both cases. Thus, phase of V_L and b_L is selected zero.

Fig. 7.2.5.1 Overview power of matching formulae

In Fig. 7.2.5.1, power matching formulae for one-ports have been summarized. All formulae can be described by one formula [Wyatt88].

References

[Baranyi84]   A. Baranyi, J. Ladvánszky, On the stability of non-linear two-port amplifiers. Circuit Theory Appl. **12**(2), 123–131 (1984)

[Chua71]   L.O. Chua, Memristor—the missing circuit element. IEEE Trans. Circuit Theory **18**(5), 507–519 (1971)

[Chua79]   L.O. Chua, C.Y. Ng, Frequency domain analysis of nonlinear systems: general theory. Electron. Circuits Syst. **3**, 165–185 (1979)

[Gelb68]   G.-V. Velde, *Multiple input describing functions and nonlinear system design* (McGraw-Hill, New York, 1968)

[Ladvánszky86A] J. Ladvánszky, On the extension of the nonlinear resistive maximum power theorem I. *Proc. of the ISCAS'86*, San José, California, 5–7 May 1986, pp. 257–259

[Ladvánszky86B] J. Ladvánszky, On the extension of the nonlinear resistive maximum power theorem II. *Proceedings of the 8th International Colloquium on Microwave Communications*, Budapest, Hungary, 25–28 August 1986, pp. 251–252

[Ladvánszky87]   J. Ladvánszky, Maximum power theorem - a describing function approach. *Proc. of the European Conference on Circuit Theory and Design*, Paris, France, 1–4 September 1987, pp. 35–40

[Mazumder77]   S.R. Mazumder, P.D. Van der Puije, An experimental method of characterizing nonlinear two-ports and its application to microwave class C transistor power amplifier design. IEEE J. Solid State Circuits **12**(5), 576–580 (1977)

[Ortega70]   J.M. Ortega, W.C. Rheinboldt, *Iterative solution of nonlinear equations in several variables* (Academic Press, New York, London, 1970)

[Korn-Korn75]   G.A. Korn, T.M. Korn. Mathematics for engineers, in Hungarian, Műszaki Könyvkiadó, Budapest, 1975

[Lin85]   P.M. Lin, Competitive power extraction from linear n-ports. IEEE Trans. Circuits Syst. **32**(2), 185–191 (1985)

[Rohrer65]   R.A. Rohrer, The scattering matrix normalized to complex n-port load networks. IEEE Trans. Circuit Theory **12**, 223–230 (1965)

[Rohrer68]   R.A. Rohrer, Optimal matching: a new approach to the matching problem for real time-invariant one-port networks. IEEE Trans. Circuit Theory **15**, 118–124 (1968)

[Wyatt83A]   J.L. Wyatt, Nonlinear dynamic maximum power theorem, with numerical method. Internal report, Massachusetts Institute of Technology, LIDS-P-1331, 1983

[Wyatt83B]   J.L. Wyatt, L.O. Chua, Nonlinear resistive maximum power theorem, with solar cell application. IEEE Trans. CAS **30**, 824–828 (1983)

[Wyatt88]   J.L. Wyatt, Nonlinear dynamic maximum power theorem. IEEE Trans Circuits Syst. **35**(5), 563–566 (1988)

Chapter 8
The Most General Solution (Wyatt)

Power extraction term can be used because we restrict ourselves to periodic excitations. Essentially the same can be repeated as for nonlinear resistive generators. The admittance operator of the optimum load [Wyatt88a] is

$$G_{\text{opt}}(v) = \left(DF_{(v)}\right)^{\text{adj}} v \qquad (8.1)$$

where D is the Gateaux differential operator [Berger77], F is the operator in the Norton equivalent of the generator, and adj is the adjoint operator:

$$\langle a, Fb \rangle = \langle F^{\text{adj}} a, b \rangle \qquad (8.2)$$

Equation (8.1) contains all that we described in the previous chapters about maximum energy transfer. However, it has theoretical significance only.

References

[Wyatt88a] J.L. Wyatt, Nonlinear dynamic maximum power theorem. IEEE Trans. CAS **35**(5), 563–566 (1988)
[Berger77] M.S. Berger, *Nonlinearity and Functional Analysis* (Academic, New York, 1977)

Chapter 9
Conclusions

After this great tour, it is time to have a rest and look back. Power matching has been followed throughout circuit theory. The summary is given in Preface.

The style of the development is that we build from simplest to more intricate. We start from the derivative-free proof of conjugate matching. Then the resistive nonlinear case is treated. Then competitive linear power matching comes and the scattering matrix. Peak of this book is the introduction of two points from foundation concepts in Chap. 6, and practically applicable solutions of the dynamic nonlinear problem in Chap. 7. Finally, we show the general solution, from which all mentioned formulae can be deduced.

In this book, we wanted to emphasize our opinion that value of an idea has been increased a lot if it is practically applicable.

We would like to point out some new results. An example is the proof for arbitrary n, that competition power matching of an n-port generator results in image parameters as loads. Another example is image parameter solution of the broadband matching problem that has been published internally some 30 years ago. A third example is the novel point of view on causality. Many other examples exist. The author sincerely hopes that the present work is useful and enjoyable.

At the time of writing these conclusions, it was recognized that many important topics have been left out. These are planned to be included in the next edition.

We are aware that more recently, some voices appeared against the maximum power transfer theorem [McLaughlin07]. These publications are based obviously on lack of knowledge. Our present eBook is a proper answer.

J. Ladvánszky, *Theory of Power Matching*, SpringerBriefs in Electrical and Computer Engineering, https://doi.org/10.1007/978-3-030-16631-1_9

Reference

[McLaughlin07]   J.C. McLaughlin, K.L. Kaiser, "Deglorifying" the maximum power transfer theorem and factors in impedance selection. IEEE Trans Educ **50**(3), 251–255 (2007)

Index

© The Author(s), under exclusive licence to Springer Nature Switzerland AG 2019
J. Ladvánszky, *Theory of Power Matching*, SpringerBriefs in Electrical and
Computer Engineering, https://doi.org/10.1007/978-3-030-16631-1